Manufacturing Execution Systems

About the Editor

Heiko Meyer has over 10 years of professional experience in developing software solutions for process and factory automation. He holds a master of science degree in mechanical engineering and a Ph.D. in computer science (modeling distributed systems) from the Technical University of Munich (Germany). At present he is head of the research and development department at Gefasoft AG. He has published over 30 papers and several books on the subject of industrial software solutions. He can be reached at heiko.meyer@gefasoft.de.

About the Contributing Authors

Franz Fuchs studied communications and electrical engineering at the University of Cooperative Education in Klagenfurt (Austria). He then worked on several projects relating to automation technology in the automotive industry. In 1984 he and two others founded Gefasoft AG (formerly GmbH). Since that time he has been a member of the executive board of Gefasoft and, among other duties, has been responsible for the basic concepts and designs of Gefasoft's MES solutions. He can be reached at franz.fuchs@gefasoft.de.

During the last 40 years, **Klaus Thiel** has worked as a consultant and project manager for several companies involved in the optimization and rationalization of the manufacturing process. He holds a master of science degree in economics from the University of Munich (Germany). In 1977 he founded Partplan GmbH. The main focuses of this company were consulting in the manufacturing environment and the development of software-based human-machine interfaces. Since 2004 he has worked as an independent production management consultant. He can be reached at info@mes-consult.de.

Manufacturing Execution Systems
Optimal Design, Planning, and Deployment

Heiko Meyer   Editor

Franz Fuchs   Contributing Author

Klaus Thiel   Contributing Author

New York Chicago San Francisco
Lisbon London Madrid Mexico City
Milan New Delhi San Juan
Seoul Singapore Sydney Toronto

The **McGraw·Hill** Companies

Library of Congress Cataloging-in-Publication Data

MES. English
    Manufacturing execution systems: optimal design, planning, and deployment/Heiko Meyer, editor; Franz Fuchs, contributing author; Klaus Thiel, contributing author.
        p.   cm.
    Includes bibliographical references and index.
    ISBN 978-0-07-162383-4 (alk. paper)
    1. Manufacturing processes—Data processing. 2. Manufacturing processes—Planning. 3. Industrial efficiency. I. Meyer, Heiko, 1971– II. Fuchs, Franz, 1961– III. Thiel, Klaus, 1940– IV. Title.
    TS183.M4613   2009
    670.285—dc22   2008055346

Copyright © 2009 by The McGraw-Hill Companies, Inc. All rights reserved. Printed in the United States of America. Except as permitted under the United States Copyright Act of 1976, no part of this publication may be reproduced or distributed in any form or by any means, or stored in a data base or retrieval system, without the prior written permission of the publisher.

1 2 3 4 5 6 7 8 9 0   DOC/DOC   0 1 5 4 3 2 1 0 9

ISBN 978-0-07-162383-4
MHID 0-07-162383-3

Sponsoring Editor
Taisuke Soda

Editing Supervisor
Stephen M. Smith

Production Supervisor
Pamela A. Pelton

Project Manager
Somya Rustagi, International Typesetting and Composition

Copy Editor
James K. Madru

Proofreader
Shivani Arora, International Typesetting and Composition

Art Director, Cover
Jeff Weeks

Composition
International Typesetting and Composition

Printed and bound by RR Donnelley.

McGraw-Hill books are available at special quantity discounts to use as premiums and sales promotions, or for use in corporate training programs. To contact a special sales representative, please visit the Contact Us page at www.mhprofessional.com.

This book is printed on acid-free paper.

Information contained in this work has been obtained by The McGraw-Hill Companies, Inc. ("McGraw-Hill") from sources believed to be reliable. However, neither McGraw-Hill nor its authors guarantee the accuracy or completeness of any information published herein, and neither McGraw-Hill nor its authors shall be responsible for any errors, omissions, or damages arising out of use of this information. This work is published with the understanding that McGraw-Hill and its authors are supplying information but are not attempting to render engineering or other professional services. If such services are required, the assistance of an appropriate professional should be sought.

This book is dedicated to my father
Dr. Karlhorst Meyer,
former professor in the Department of Mathematics,
College of Liberal Arts and Sciences,
University of Florida, Gainesville

HM

Contents

Foreword		xv
Acronyms		xvii

1 Introduction ... 1
  1.1 Motivation ... 1
  1.2 Aim of This Book ... 2
  1.3 Structure of This Book ... 3

2 Factory of the Future ... 5
  2.1 Historical Development of Manufacturing Execution Systems ... 5
    2.1.1 Development of Business Data Processing ... 5
    2.1.2 The Integration Concept: From CIM to the Digital Factory ... 6
  2.2 Definitions of Terms ... 8
    2.2.1 Classification of Terms ... 8
    2.2.2 Company Management Level ... 8
    2.2.3 Production Management Level ... 10
    2.2.4 Control/Automation Level ... 13
  2.3 Shortfalls of Existing Architectures and Solutions ... 13
    2.3.1 Patchwork ... 13
    2.3.2 No Common Database ... 14
    2.3.3 Excessive Response Times ... 14
    2.3.4 High Operating and Management Outlay ... 15
  2.4 Demands of Future Production Management Systems ... 16
    2.4.1 Target Management ... 16
    2.4.2 Integration of Applications and Data ... 18
    2.4.3 Real-Time Data Management ... 21
    2.4.4 Information Management ... 24
    2.4.5 Compliance Management ... 25
    2.4.6 Lean Sigma and MES ... 27
  2.5 Summary ... 30

3 Concepts and Technologies ... 31
  3.1 Commonalities between Existing Approaches and MES ... 31

Contents

- 3.2 Norms and Guidelines 31
  - 3.2.1 ISA 31
  - 3.2.2 IEC 35
  - 3.2.3 VDI 36
  - 3.2.4 FDA 36
  - 3.2.5 NAMUR 37
- 3.3 Recommendations 38
  - 3.3.1 MESA 38
  - 3.3.2 VDA 39
  - 3.3.3 VDMA 40
  - 3.3.4 ZVEI 40
- 3.4 Adjacent Areas 41
  - 3.4.1 Historical Development of ERP/PPS Systems 41
  - 3.4.2 ERP/PPS Systems 41
  - 3.4.3 Process Management Systems 42
  - 3.4.4 SCADA Systems 45
  - 3.4.5 Simulation Systems 45
- 3.5 Product Lifecycle Management 46
  - 3.5.1 Historical Development 46
  - 3.5.2 Product Model 47
  - 3.5.3 Process Model 48
  - 3.5.4 Implementation Strategies 48
  - 3.5.5 Points of Contact with MES 49
- 3.6 Summary 50

4 Core Function—Production Flow-Oriented Design 53
- 4.1 Cross-System Cohesiveness 53
  - 4.1.1 Classification in the Overall System ... 53
  - 4.1.2 General and Complete Data Model ... 54
  - 4.1.3 Origins of Master Data 56
- 4.2 Data Model for Product Definition 57
  - 4.2.1 Relevant Concepts 57
  - 4.2.2 The Operation 59
  - 4.2.3 The Work Plan 63
  - 4.2.4 The Parts List 66
  - 4.2.5 Change Management and Product History 66
- 4.3 Data Model for Resource Management 66
  - 4.3.1 Description of Production Environment 66
  - 4.3.2 Production Personnel 71
  - 4.3.3 Operating Resources 72
  - 4.3.4 Materials and Preliminary Products ... 73
  - 4.3.5 Information and Documents 74

4.4	System and Auxiliary Data		77
4.5	Order Fulfillment Data		79
	4.5.1	Orders	79
	4.5.2	Production Data, Operating Data, and Machine Data	79
	4.5.3	Derived Performance Data and Figures	81
4.6	Summary		81

5 Core Function—Production Flow-Oriented Planning ... 83

5.1	Integration within the Overall Process		83
5.2	Order Data Management		83
5.3	Supply Management within the MES		85
	5.3.1	Demand Planning	85
	5.3.2	Material Requirement Calculation	86
	5.3.3	Material Disposition in the MES or ERP System	86
	5.3.4	Incoming Goods	87
	5.3.5	Interaction between the ERP System and the MES	87
	5.3.6	Material Warehousing Costs	88
5.4	The Planning Process		88
	5.4.1	Planning Objectives	88
	5.4.2	The "Updated" Work Plan: Condition for Optimized Planning	89
	5.4.3	Work Scheduling	89
	5.4.4	Strategies for Sequence Planning and Planning Algorithms	91
	5.4.5	Forward Planning/Reverse Planning/Bottleneck Planning	92
	5.4.6	Collision-Free Planning of a Time Container	93
	5.4.7	Setup Optimization and Warehousing Costs	94
5.5	The Importance of the Control Station		94
	5.5.1	Core Elements	94
	5.5.2	User Interface	95
5.6	Personnel Planning and Release of Orders		96
5.7	Summary		97

6 Core Function—Order Processing ... 99

6.1	General Information on Order Processing		99
	6.1.1	Classification within the Overall System	99

Contents

- 6.1.2 Functions of Order Fulfillment 99
- 6.1.3 The MES Terminal 100
- 6.2 Order Preparation and Setup 103
  - 6.2.1 Changing Tools 103
  - 6.2.2 Machine Settings 104
  - 6.2.3 Material Provision 105
  - 6.2.4 Test Run 105
- 6.3 Order Control 106
  - 6.3.1 Information Management 106
  - 6.3.2 Control and Tracing of Production Units 106
  - 6.3.3 Managing the Production Bin 106
  - 6.3.4 Material Flow Control 106
  - 6.3.5 Order Processing and Operating Data Recording 109
  - 6.3.6 Process and Quality Assurance 110
- 6.4 Performance Data 112
  - 6.4.1 Involved Departments 112
  - 6.4.2 Key Figures and Performance Record 116
  - 6.4.3 Ongoing Analysis and Evaluations 116
  - 6.4.4 More Long-Term Analyses and Evaluations 117
- 6.5 Maintenance Management 118
  - 6.5.1 Tasks 118
  - 6.5.2 Preventive Maintenance and Repair 118
  - 6.5.3 Alarm Management 119
- 6.6 Summary 119

7 Technical Aspects 121
- 7.1 Software Architecture 121
  - 7.1.1 Fundamental Variants 121
  - 7.1.2 Overview of Central Components 122
  - 7.1.3 Platform Independence 124
  - 7.1.4 Scalability 125
  - 7.1.5 Flexible Adjustment versus Suitability for Updates 127
  - 7.1.6 MES and Service-Oriented Architecture (SOA) 129
- 7.2 Database 130
  - 7.2.1 Introduction 130
  - 7.2.2 Resource Monitoring 130
  - 7.2.3 Scaling the Database System 132
  - 7.2.4 Data Management and Archiving 133
  - 7.2.5 Running Maintenance 134

	7.3	Interfaces with Other IT Systems	134
		7.3.1 Overview	134
		7.3.2 Interface with Production	134
		7.3.3 Interface with an ERP System	140
		7.3.4 Interface with the IT Infrastructure	143
		7.3.5 Interface with Communication Systems	144
		7.3.6 Other Interfaces	144
	7.4	User Interfaces	144
		7.4.1 Usage and Visualization	144
		7.4.2 Reporting	148
		7.4.3 Automated Information Distribution	150
	7.5	Summary	151
8	**Evaluation of the Cost-Effectiveness of MES**		**153**
	8.1	General Information on Cost-Effectiveness	153
		8.1.1 Calculation of Cost-Effectiveness	153
		8.1.2 Comparative Cost Method	154
		8.1.3 Value-Benefit Analysis	154
		8.1.4 Performance Measurement	155
		8.1.5 Total Cost of Ownership	155
	8.2	General Information on Evaluation	156
		8.2.1 Assessing Cost-Effectiveness in Practice	156
		8.2.2 Rationalization Measures in Production	157
		8.2.3 MES for Reducing Sources of Loss	160
	8.3	The Benefits of an MES	161
		8.3.1 Integrated Data Transparency	161
		8.3.2 Reducing Time Usage	162
		8.3.3 Reducing Administration Expenses	165
		8.3.4 Improved Customer Service	166
		8.3.5 Improved Quality	166
		8.3.6 Early Warning System, Real-Time Cost Control	167
		8.3.7 Increasing Employee Productivity	167
		8.3.8 Compliance with Directives	168
	8.4	The Costs of an MES	168
	8.5	Summary	169
9	**Implementing an MES in Production**		**171**
	9.1	Implementing IT Systems in General	171
		9.1.1 Selection of Components	171
		9.1.2 Implementation Strategies	173
		9.1.3 Problems during Implementation	174

Contents

- 9.2 Preparation of the Implementation Project ... 176
  - 9.2.1 Establishing the Core Team ... 176
  - 9.2.2 The Fundamental Decision: MES: Yes or No ... 177
  - 9.2.3 Establishing the Project Team ... 177
- 9.3 Analysis of the Actual Situation ... 178
  - 9.3.1 Introduction ... 178
  - 9.3.2 Existing Infrastructure ... 179
  - 9.3.3 Existing Processes and Required Functions ... 180
  - 9.3.4 Key Figures as the Basis for Monitoring Success ... 181
  - 9.3.5 Suitable Key Figures for Success Monitoring ... 182
  - 9.3.6 Other Factors for Success ... 184
- 9.4 Creation of a Project Plan ... 185
- 9.5 Contract Specifications ... 186
- 9.6 Selection of a Suitable System ... 187
  - 9.6.1 Market Situation ... 187
  - 9.6.2 Short-Listing and Limiting to Two or Three Applicants ... 187
  - 9.6.3 Detailed Analysis of the Favorites and Decision ... 189
- 9.7 Implementation Process ... 191
  - 9.7.1 Project Management ... 191
  - 9.7.2 Training Management ... 192
  - 9.7.3 Operating Concept ... 193
- 9.8 Summary ... 195

10 Examples for Application ... 197
- 10.1 Mixed Processes ... 197
- 10.2 Sensient Technologies: Emulsions ... 198
  - 10.2.1 Information on Sensient Technologies Corporation ... 198
  - 10.2.2 Description of the Production Process ... 198
  - 10.2.3 Basic Quantity Units and Production Units ... 200
  - 10.2.4 Production Process Plan ... 200
  - 10.2.5 Challenges for the MES ... 200
  - 10.2.6 Realization and Implementation ... 204
- 10.3 Acker: Synthetic Fiber Fabrics ... 204
  - 10.3.1 Information on the Company ... 204
  - 10.3.2 Description of the Production Process ... 204

		10.3.3	Basic Quantity Units and Production Units	209
		10.3.4	Production Flow Plan	210
		10.3.5	Tasks of the MES	210
		10.3.6	Challenges	212
		10.3.7	Realization and Implementation	213
	10.4	Summary		214
11	Visions			215
	11.1	Merging the Systems		215
	11.2	The MES as a Medium of Product-Development Management		217
		11.2.1	Phases of Product Development	217
		11.2.2	Request Handling	217
		11.2.3	Concept Documentation and Designing Requirements	218
		11.2.4	Construction of the Product	219
		11.2.5	Computer-Aided Flow Planning	220
		11.2.6	Production Management	221
	11.3	Standardization of Function Modules		221
	11.4	Merging Consultancy Activities and IT Systems		221
	11.5	Summary		222
12	Summary of the Book			223
	References			227
	Glossary			229
	Index			245

Foreword

When a general business concept becomes absorbed into the mainstream, it sometimes loses some of the sharpness associated with its original formulation. This is a charge that may be leveled with some justification against enterprise resource planning (ERP). On the one hand, an ERP system promises—by the force of its title—to link the entire enterprise together in a comprehensive resource plan. However, ERP systems in real life are far less ambitious. They are equivalent to software to automate a firm's accounting and administrative systems. This narrowness of outlook has greatly hindered the vital overlap between a firm's ERP system and the system governing the automation of its production planning and execution functions. The latter system is the firm's manufacturing execution system (MES). The aim of the current book is to spell out in detail the design of an MES. Along the way, the tricky question of how the ERP system should interface with the MES is clearly answered.

This book is the clearest exposition I have seen of the ideal anatomy of a production-oriented IT system. The fundamentals of product mapping, operations sequence planning, and production control with material management, data management, maintenance management, and quality management are lucidly explained. The authors take great care to avoid clouding concepts with unnecessary jargon: every piece of terminology is carefully and precisely defined before it is put to use.

An introductory chapter and two additional chapters describe the backdrop of MES. These three chapters will serve as a useful review for IT professionals and academics and as a valuable primer for non-IT industry professionals and academics. The meat of the book is in Chaps. 4 through 6, which explain the core functions of MES (production flow-oriented design, production flow-oriented planning, and order execution). The following three chapters cover technical aspects of MES (including interfaces to ERP and to IT infrastructure). However—perhaps biased by the fact that I am an operations management academic—one of my favorite chapters in the book is Chap. 10. Here the reader will find two valuable examples of production systems viewed through the MES lens filled out in great detail. One application is drawn from the continuous process industry; the second example is from the apparel industry and has the dual character of discrete parts manufacturing and batch production. The authors describe each process in detail and list specific challenges that an MES for each process would have to address.

One of the merits of this book is that it is written from first principles and will therefore be found palatable to decision makers within an organization (such as financial managers and production managers) in addition to IT professionals. Academics in IT and operations management will find valuable material for courses at the interface of IT and manufacturing management. It is a challenging task to write a book with an IT theme that non-IT professionals will find useful and interesting. This book is one such.

Dr. Anand Paul, Associate Professor
Department of Information Systems and Operations Management
Warrington College of Business Administration
University of Florida, Gainesville

Acronyms

ABC	Activity-based costing
ADO	ActiveX data objects
AJAX	Asynchronous JavaScript and XML
ANSI	American National Standards Institute
API	Application programming interface
APS	Advanced planning and scheduling
ATP	Available to promise
CAD	Computer-aided design
CAE	Computer-aided engineering
CAM	Computer-aided manufacturing
CAP	Computer-aided process planning
CAQ	Computer-aided quality assurance
CI	Corporate identity
CIM	Computer-integrated manufacturing
COM	Component Object Model
Cpk	Process capability index
CPM	Collaborative production management
CRM	Customer relationship management
CSV	Comma/character-separated value
DBMS	Database management system
DCOM	Distributed Component Object Model
DMAIC	Define, measure, analyze, improve, and control
DMU	Digital mock-up
DNC	Distributed numerical control
EDM	Engineering data management
EDP	Electronic data processing
EJB	Enterprise JavaBeans
EPM	Enterprise production management
ERP	Enterprise resource planning
FDA	Food and Drug Administration
FIM	Finance management
FPY	First-pass yield
GSD	General station description
HMI	Human-machine interface
HRM	Human resources management

HTTP	Hypertext Transfer Protocol
ID	Identifier
IPC	Industrial PC
IT	Information technology
KPI	Key performance indicator
LCD	Liquid-crystal display
LDAP	Lightweight Directory Access Protocol
LED	Light-emitting diode
LIMS	Laboratory information management system
MDA	Machine data acquisition
MDM	Master data management
MES	Manufacturing execution system
MRP	Material resource planning
NC	Numerical control
ODBC	Open database connectivity
OEE	Overall equipment efficiency
OLAP	Online analytical processing
OLE	Object linking and embedding
OPC	OLE for process control
PAA	Part average analyses
PCS	Process control system
PDA	Production data acquisition
PDM	Product data management
PLC	Programmable logic controller
PLM	Product lifecycle management
Ppk	Process performance index
PPS	Production planning and scheduling
RFC	Remote function call
RFID	Radio frequency identification
ROI	Return on investment
RPC	Remote procedure call
SCADA	Supervisory control and data acquisition
SCM	Supply-chain management
6σ	Six Sigma/6Sigma
SMED	Single-minute exchange of die
SMS	Short Message Service
SNMP	Simple Network Management Protocol
SOA	Service-oriented architecture
SOAP	Simple Object Access Protocol
SPC	Statistical process control
SQC	Statistical quality control
SQL	Structured Query Language
SRM	Supplier relationship management
STEP	Standard for the Exchange of Product Data
SVG	Scalable vector graphics
SWF	Shock Wave Flash
TCO	Total cost of ownership

TCP/IP	Transmission Control Protocol/Internet Protocol
TPM	Total productive maintenance
TPS	Toyota production system
TQM	Total quality management
UDDI	Universal description discovery and integration
W3C	World Wide Web Consortium
WIP	Work in process
WPF	Windows Presentation Foundation
WSDL	Web Service Description Language
XAML	eXtensible Application Markup Language
XML	eXtensible Markup Language

Manufacturing Execution Systems

CHAPTER 1
Introduction

1.1 Motivation

The globalization of the economy and the associated factors of increasing effectiveness in production, shortening innovation cycles, safeguarding high quality, etc. are continually augmenting the pressure on the production business. It has been possible to compensate somewhat for this pressure in recent years by relocating production to low-cost countries. However, in the medium term, the demands of workers in countries that are still low cost will increase, and production costs will rise as a result, so the need for action will arise. Tools will be needed to increase efficiency in existing production processes. It also must be considered that production in high-cost countries definitely has its advantages, so these countries are becoming more and more feasible as production locations and will remain so in the long term. The degree of automation is already extremely high in these countries, so modifying production processes will not increase efficiency significantly.

Additional new challenges for production-oriented information technology (IT) systems arise from norms and guidelines, such as quality assurance standards and regulations in the food and pharmaceutical industries. While these demands were relevant mainly for security-oriented systems decades ago, now transparency and traceability are playing an increasingly important role in other sectors as well.

In order to achieve effective value creation in production, equipment is needed that can meet these new demands 100 percent. Existing enterprise resource planning (ERP) systems established on the market are largely administrative and accounting systems. The new systems needed must include functions for planning, logging, and control that not only act but also react in real time. For these systems, the concept of a *manufacturing execution system* (MES) has arisen. Since MES is a multifaceted area, each sector interprets the concept from its own standpoint. In addition, there are various software suppliers on the market who are offering their systems—although misleadingly—as MES.

An alternatively used and more meaningful term for MES is *collaborative production management* (CPM), but this term thus far has not become well established. Both new IT systems are ultimately concerned with controlling production using integrated information and monitoring equipment in order to achieve predefined target values. These targets include real-time performance indexes and trend analyses for internal processes as well as figures on the quality of the company in the perception processes of society. This involves verification of compliance with directives and, recently, voluntary proof regarding target achievement indexes in the social and ecological sectors as well.

The concept of *collaborative* that is contained in CPM is meant to indicate that it is not just the core elements such as planning, implementation, and the recording and examination of information that work together, but that the peripheral areas such as ERP, marketing, and purchasing are also involved in the exchange of information. With the aid of suitable Web technologies, these systems form an *enterprise production management* (EPM) system, in which MES data and information are made available to all those in the plant involved in the value-creation process in an event-based form.

1.2 Aim of This Book

This book is aimed primarily at decision makers within a company, such as managing directors, financial managers, controllers, and production managers, who are affected by the implementation of an MES. Second, because of its neutral representation of the concept of MES, this book can be used as an introduction to production-oriented software systems.

On one hand, the content of an MES is described in detail in a neutral form and avoiding the use of too many terms so that the topic can be handled with more transparency. This book examines the *ideal concepts* of MES, which also could be the basis for new developments of relevant systems. On the other hand, the aim is to show that these systems, in contrast to ERP systems, include huge advantages that could, with proper implementation, far outweigh the costs connected with them. Generally, decision makers are rather conservative when purchasing software systems. Skepticism is understandable because purchasing, implementation, and training costs are not insubstantial. We therefore demonstrate that it is possible to measure and prove the advantages of these systems using relatively simple means before implementing them.

As already suggested in Sec. 1.1, there are also would-be MES on the market that make it difficult for potential users to select a suitable system. In addition, existing systems differ significantly in their suitability, functionality, and potential areas of application. Ultimately, management must select a system. This book is intended to provide

Introduction 3

the relevant basic knowledge about MES in a neutral way in order to make it possible to compare the systems on the market and assess them based on their suitability.

1.3 Structure of This Book

The structure of this book is represented schematically in Fig. 1.1. The book consists of 12 chapters in all, which are arranged as follows:

After a short introduction, Chap. 2 will examine the requirements of the factory of the future. The necessary characteristics for a production-oriented solution will be described in the process. These can be realized only through a central, strategic IT system with a uniform, consistent data model. Before an approach is developed for the requirements identified in Chap. 2 for the factory of the future, the merits of existing standards and technologies that can be drawn on to solve the problem are examined in Chap. 3.

After an analysis of the requirements and existing approaches, the following three chapters describe the core functions of MES.

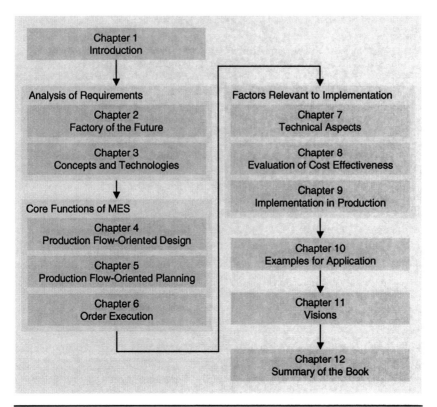

FIGURE 1.1 Structure of this book.

Chapter 4 examines production flow-oriented design. Chapter 5 addresses the tasks related to production flow-oriented sequence planning. Finally, Chap. 6 examines order fulfillment.

However, in addition to the theoretical fundamentals of MES, this book is also intended to provide the reader with concrete aids for implementing MES. Thus the technical aspects, such as the necessary hardware, integration of the control level, etc., will be looked at in more detail in Chap. 7. Chapter 8 shows the ways in which the cost efficiency of MES can be assessed. One very important topic with regard to the purchase of complex software systems is ensuring acceptance among employees and conducting systematic training. Chapter 9 provides helpful methods for the implementation of relevant systems. Your projects' prospects for success should be increased considerably by these strategies.

Two concrete examples from production and process technology for the implementation of MES are given in Chap. 10. After a description of the production processes, we look at the challenges in concrete cases and look briefly at realization and implementation.

Chapter 11 examines the outlook for the future and describes the authors' visions of future production-oriented IT systems. It is predicted that the various existing software solutions will merge and grow together.

Chapter 12 provides a summary of the entire book and concludes with a summary of further anticipated developments in MES.

CHAPTER 2
Factory of the Future

2.1 Historical Development of Manufacturing Execution Systems

2.1.1 Development of Business Data Processing

In order to better understand the challenges to information technology (IT) in the factory of the future, we need to take a glance at the development of business data processing. This accompanied the rapidly growing need for information on the part of market participants.

In the 1970s, mainframe computers, which generally served a large number of users in computer centers, ruled the computer world. The main issue for companies was the conversion from manually maintained accounting systems to electronic systems. As supplements to accounting, order-processing systems with the central function of invoicing and purchasing systems (including material administration) evolved. In parallel, independent personnel administration systems with integrated payroll functions were developed.

Only in the last-third of the 1970s did minicomputers that were affordable for medium-sized companies become available. An ever-increasing number of software suppliers are still developing programs for the aforementioned core tasks.

It was also in this period that *material requirement planning* (MRP) systems came into being, although these systems hardly did justice to this title. However, the first software products for production did perform important partial tasks for an MRP system (from today's point of view, partial functions of a *manufacturing execution system* (MES):

- *Production data acquisition (PDA) and machine data acquisition (MDA).* Data were acquired manually using nonintelligent data-capture devices. With the transition from conventional contactor controls to programmable logic controllers (PLC), the transfer of production and machine data could be automated more and more, and as a result, the data were more recent and of a better quality.

- *Computer-aided design (CAD)*. The beginning of the 1980s marked the birth of electronic drawing systems, which led to huge increases in productivity in construction and design. Here, a revolutionary development has taken place up to today: three-dimensional (3D) CAD replaced two-dimensional (2D) CAD systems. In electronic engineering, more and more powerful product-development tools (for finite-element computation, simulation, etc.) have been produced. The standard achieved today offers systems that simulate, operate, and visualize the product virtually (DMU = digital mock-up). These systems are at the core of the product lifecycle management (PLM) concept with which the lifecycle of a product is recorded (see Sec. 3.5).

- *Computer-aided quality assurance* (CAQ). Also at the start of the 1980s, development began on quality assurance systems as a result of increasingly precise quality standards for products. This went hand in hand with the development of software for statistically ensuring the process capability of products (SPC = statistical process control). Within the framework of integration concepts, quality assurance is currently experiencing a renaissance.

2.1.2 The Integration Concept: From CIM to the Digital Factory

Overview
All these systems were isolated solutions for a specific department. As a result, the idea of developing integrated production systems (CIM = computer-integrated manufacturing) came into being in the mid-1980s. Although the task was recognized early on, implementation often failed owing to complexity because neither the notion of standardization nor the technologies available had been developed sufficiently. The willingness to provide the required investment also was lacking to some extent.

Only through globalization with the consequence of increasingly tough competition did companies come under previously unknown pressure to be efficient. They faced challenges to be faster, better, cheaper. The need for integrated observation of the performance processes of production in real time was quickly recognized. Altered demands on production management systems and the information to be provided were closely connected with this.

The factory of the future systematically uses further developments in IT. Even before the actual production process, the factory is virtually mapped out as holistically and true to detail as possible.

The optimal production process for the product is developed interactively through simulation of the production process, including the material flow and information exchange. One related concept is the *digital factory*. Behind this striking concept hides an integrated data model of the future production process in which every planning and production process is illustrated cohesively. For the digital factory, modern methods and software tools are applied in order to test systems and production processes in extensive simulations. It thus should be possible to ensure long before the start of actual production that every step is coordinated and that all facilities run smoothly. Typical startup times of 3 to 12 months can be greatly reduced in this manner. The manufacturers also expect a reduction in the planning phase and an improvement in the quality of the product.

The digital factory thus is a counterpart of the virtual (digital) product, whereby the virtual product and digital factory should arise in parallel (→ short product cycles at reduced planning costs) and are closely interconnected with each other. In virtual processes, production is simulated, evaluated, and after a successful cycle is released for the actual process. Some important components for achieving these ambitious goals are listed below.

Standardization

Methods, processes, and resources should be standardized to such an extent that they can be used again for a new-product or follow-up model with the fewest possible changes (ideally, with no changes) in accordance with a building-block principle. Here, it should be systematically calculated whether or not it makes sense to use the latest technology or whether you should stick to the tried and tested for the benefit of standardization. This not only reduces costs when purchasing parts and facilities but also offers significant advantages with regard to maintenance, flexibility, and reliability. The standardization of production systems (i.e., communication, control engineering, etc.) is also one of the most important conditions for the efficient implementation of an MES.

Data Integration

All relevant planning data (on the product, on the process, and on the resources) are captured only once by the persons concerned and are managed centrally in a common database. Thus they are always available in the latest format for every planner and increasingly for suppliers and service providers as well. Another goal is to use these data in the planning of new products in order to obtain sound costing as soon as possible.

A deciding factor here is a *central and consistent database* in which all product-relevant data can be managed by the various responsible divisions and then, with the aid of a detailed permission management, be made available to users in planning and production across all company sites. The further development of this integration concept results in the concept of PLM (see Sec. 3.5). In these systems, data arising after production are also captured; that is, the entire history of a product is illustrated "from the cradle to the grave" through data links. The data and their contents then are available for the development and production of new products and speed up the development process. What is "new" about this is that the development process accesses the entire data pool of similar products and is not restricted to 3D visualizations alone, but suggests the entire data framework for the actual process and coordinates and adjusts it to production in real time. Here, it is also important that the results of actual processes and information from change management are incorporated into the development once again.

Automation in Engineering
Software tools handle many aspects and routine activities automatically. Alternatives are even generated automatically through the variation of individual parameters. However, a condition for this is extensive standardization—especially in the area of production facilities (machines and related controls).

Process and Change Management
All those involved in the process carry out their tasks on the basis of defined processes (workflows) with extensive support from IT systems. This ensures the availability of the required data at the right time, in the right level of detail, and in the right context.

2.2 Definitions of Terms

2.2.1 Classification of Terms
The world of IT—and especially the subject of production systems—is marked by a confusing abundance of terms. Standardization bodies or user organizations defined many of these concepts; others simply were derived from industry- or company-specific concepts and generalized.

The overview shown in Fig. 2.1 contains just a small selection of these terms that have become part of everyday language usage and that are also used in this book. In order to clarify the classification of the terms in the Instrumentation, Systems, and Automation Society (ISA) level model, they are also displayed in the form of a diagram.

2.2.2 Company Management Level
The company management level comprises the central functions of accounting and finance, sales and marketing, purchasing, and human resources management and generally is represented by an *enterprise*

Factory of the Future 9

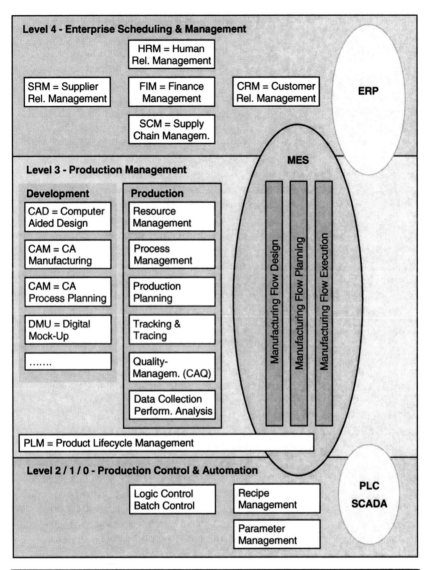

FIGURE 2.1 Overview of terms with regard to the ISA level model (see also [ISA S95-1]).

resource planning (ERP) system, in which these functions are mapped. An ERP system generally also contains planning functions for evaluation of the medium- and long-term material requirements and for liquidity planning. The core function of an ERP system is *finance management* (FIM), in which the cash flows of the company are managed and monitored.

The sales and marketing function is closely connected with it and is a central component of a *customer relationship management* (CRM) system. In these systems, all relational customer data are managed and analyzed. This includes inquiries, quotations, orders, sales analyses, medium and long-term demand forecasts, and marketing functions. A CRM tool can be used as an independent function at the company management level or as an integrated component of the ERP system. There are also points of contact between CRM systems and MES. Delivery dates for customer orders will no longer be estimated by sales and marketing in the future but will be calculated "exactly" by the operative planning tool within the MES. The MES thus is required for improving adherence to delivery dates (and thus customer satisfaction) in the sense of the "capable to promise" concept.

Another main task at the company management level is purchasing. As a counterpart to CRM tools, a supporting *supplier relationship management* (SRM) system is often used here. In these systems, all relational data for suppliers are managed and analyzed. This includes supply contracts with agreed prices, quality levels, and delivery dates. The operative tasks of material administration, such as ordering processes, are carried out by the MES within the operative production planning and sequence establishment.

The management of employees and payroll processes also constitutes tasks at the company management level. The functions of *human resources management* (HRM) are usually mapped as a function module of the ERP system. Interfaces to the MES arise especially in the area of recording employee working hours and operating data in case, for example, the MES must provide the quantity produced per time unit and employee or per group for a piece-rate pay model or a group piece-rate pay model.

In the context of globalization, global logistics (SCM = supply-chain management) is becoming more and more important. SCM systems are integrated into ERP systems or represent an independent function at the company management level. They make it possible to better plan and monitor the logistics process in the face of global competition. In contrast to them are internal logistics processes that are part of an MES; that is, similar functions are mapped in both systems, but at different levels and with different objects.

2.2.3 Production Management Level

The production management level includes product development and actual production. There are a large number of supporting tools for product development, from simple CAD to DMU. For the production process, there is MES. Here, it is intended to provide a summary of all possible aspects (and therefore functions), which are known by many names, including the following:

- Production system
- Production control system

- Production management system (PMS)
- Product information system
- Production information system (PIS)
- Productions data system (PDS)
- Product data system
- Production data acquisition (PDA)
- Production planning system (PPS)
- Employee information system (EIS)
- Paperless manufacturing
- Electronic product data sheet
- Control system
- Control center
- Fine planning system, etc.

At the end of the 1990s, the need for better and faster product information systems became apparent. It was believed initially that an independent production management level would be rendered superfluous by integrating the automation level in ERP. The results were rather modest. This is also understandable because the reliable but cumbersome management and settlement level does not suit the real-time results-oriented world of production. The dissatisfaction of production with regard to real-time information in a large number of production companies then led in steps to preparation of guidelines for production. The ISA with its guidelines is one such example. The term *manufacturing execution system* (MES) goes back to the 11 functions for a production system developed by Manufacturing Enterprise Solution Association (MESA) at the beginning of the 1990s. The ISA acted on these recommendations and modified or extended them systematically into guidelines for batch processes (S88 Standard) and general processes (SP95).

With the creation of the term *manufacturing execution system*, this muddle of concepts could be simplified to some extent. An MES must have the entire production process under control and therefore must cover all aspects from the preceding list. This requires mapping the following functions in an integrated MES:

- The complete technical description of the product ("product definition management") and its management. The task schedule is a central component.
- Management of all resources required for the product ("resource management") and their allocation into the task schedule.
- Planning the order pool and establishing a sequence.

12   Chapter Two

- Integrated performance recording and performance monitoring ("tracking and tracing").
- Documenting performance data for traceability of production data ("traceability") and in order to meet verification guidelines.
- Information management.

The MES must map these functions in the form of software processes. Corresponding to the operating cycles of real production, a *flow-oriented approach* is selected here, in which the individual functions are allocated to three workflows in the form of adequate software processes (Fig. 2.2):

1. *Production flow-oriented design.* The production process (at the core of the task schedule in all details) is mapped with user-friendly graphic (where possible) planning tools. This production flow is saved as a data model, which should be as complete and consistent as possible, for all articles to be produced.

2. *Production flow-oriented planning.* The production flow is planned in the form of consecutive production orders (function "production planning"). The flow-oriented approach extends not only to the sequence of the orders, but also to all necessary resources, with a solution at the operating sequence level. The provision of human resources, equipment, raw materials, or components thus is also planned.

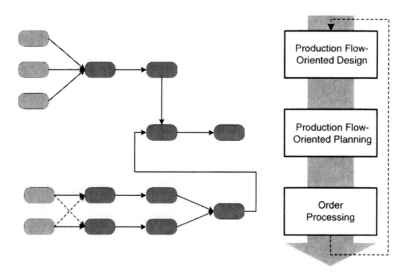

Figure 2.2   Software processes (workflows) of a flow-oriented MES.

3. *Order processing.* All other functions of the MES mentioned earlier reside in the actual implementation process. The production process is controlled and monitored, and the resulting product and production-relevant data are collected and documented in a flow-oriented manner.

The classification of the processes into these three main processes of actual production is a conceptual approach that is also used in all subsequent chapters of this book (see Chaps. 4 through 6).

To accompany the entire lifecycle of a product from the development process through the production process to the service process (as well as beyond to the scrapping of the product), there is now the term *product lifecycle management* (PLM; see also Sec. 3.5), which, however, maps only the core construction process. For processes going beyond this (e.g., enquiry process, CAM, and CAP), there are separate systems. MES today is focused on the actual production process and imports data from the development process. In the course of a product's "life," all changes are documented and coordinated with development. MES is the decisive integration platform for this.

2.2.4 Control/Automation Level

This level is controlled in automated production by *programmable logic controller* (PLC) systems and robots. The degree of automation of production depends on the quantities produced and the complexity of the tasks. Generally, manual and partly automatic workstations can be found in production as well as fully automatic stations.

Different demands grew out of this nonhomogeneous environment for MES. For automated areas, suitable mechanisms for data exchange must be provided. But for stations where humans do a large portion of the work, user-friendly operating interfaces must be available.

Often, *supervisory control and data acquisition* (SCADA) systems are used at the control level, especially in complex machines and workstations. These systems generally take on some functions of an MES, such as the management of recipes or machine parameters. In these cases, the MES acquires additional "contacts." In order to avoid double maintenance efforts and to guarantee data security in a central system, all relevant data from the SCADA systems should be maintained via the MES.

2.3 Shortfalls of Existing Architectures and Solutions

2.3.1 Patchwork

The concept *patchwork* has a positive connotation in its original meaning, describing an artistically arranged, harmonious collection of different colors and materials. Here, it refers to the use of various software tools and components together, where these are *not* coordinated with each other and therefore do not yield a positive overall effect.

The reasons as to why such scenarios arise are generally similar: for an urgent task (e.g., machine data capture), a system is installed that is tailored to exactly that task. In parallel, a similar system arises in another area (e.g., for collecting order data and managing recipes). After some time, software tools become established in many areas of the company fulfilling similar and overlapping tasks. The costs for maintaining these systems are enormous. In addition, a great deal of master data are managed at several different points, and the consistency of these data is difficult to ensure. Management is also unable to base any decisions on this patchwork because it is not possible to make across-the-board analyses of the data set without incurring unreasonable expenses.

However, there is a light at the end of the tunnel—the *patchwork principle* also has its advantages. Behind every system there are one or more persons who have implemented the system in their sectors and who therefore identify with the system and work continually to improve it. This means that the system is "alive" and is used in the best way possible. This is not always the case for a system that has been implemented centrally (see Chap. 9).

2.3.2 No Common Database

All parts of the production system need a specific database. Although a large part of the master data required is already present in the ERP system, it is not always present in the degree of detail required by the MES. As a result, discussions arise time and again regarding where the product data should be managed and who is to be responsible for them.

Independent of this subject, various databases are often found within "one" MES. Here, it is generally a case of patchwork (see Sec. 2.3.1), which appears to be an integrated product only through the cloak of a common user interface.

The concepts of the *data warehouse* and *master data management* (MDM) offer one way out of this dilemma. Here, the various systems and models of an MES have access to a common database. Double data management and inconsistencies thus can be largely avoided. However, these concepts are associated with increased technical and organizational outlay.

2.3.3 Excessive Response Times

Shortening all processes in the order-production-delivery cycle is a goal of all producing companies. In order to meet this challenge, certain sales and purchasing functions move from the (sluggish) company management level to the (rapid-response) production management level. In sales and marketing, this includes inquiry, order, delivery date, and order completion data. In purchasing, it involves the production of raw materials and half-products. However, these demands (and hopes) are only partially fulfilled by today's MES. More and more complex demands from a functional and IT viewpoint have not improved the response times of these systems. Here, it is necessary to concentrate on the core tasks once more (What does my production actually need?).

One aspect with regard to response times and "real-time behavior" is the ongoing monitoring of costs. An *activity-based costing* (ABC; see Sec. 2.4.3) integrated into the MES can speed up decision processes that today often do not take place until "final costing" with a considerably delay, that is, too late.

2.3.4 High Operating and Management Outlay

As well as the investment and implementation costs, the costs for ongoing operation of the system stand in opposition to the potential for rationalization and savings that are opened up by introducing an MES. It is essential to minimize this often considerable outlay for the running and management of an MES with appropriate software architecture:

- *Updates and release management.* The system generally must be capable of being updated and may not contain any customer- or project-specific data elements (this contradicts the demand for highly flexible tailor-made systems). It must be possible to carry out updates within a manageable period of time and without manual adjustment of the existing applications. New releases should be issued about once or twice a year. It also must be possible to skip releases (i.e., to forego updates).
- *User administration.* The authentication of the user should be possible with a central (in many cases, already existing) system. It is not practical to maintain users and passwords in parallel systems.
- *IT integration.* It is well known that the only constant in production is change. In this sense, the interfaces of the MES often need to be adjusted to suit altered demands. Here, either solutions based on script programs (which can be changed by trained key users of the customer) or service-oriented approaches (SOA = service-oriented architecture) are necessary. Permanently implemented and fixed interfaces drive up costs in the operating phase.
- *Licensing model.* Most well-known licensing models are based on counting users or workstations. This means that license costs are directly related to the number of users. This approach

may be fair and relatively easy to monitor (from the system provider's point of view), but it is certain to prevent the system from spreading freely to all areas of production and therefore cannot achieve optimal effectiveness. A better way would be to scale the licensing costs across the production resources (e.g., the existing machines and workstations).

2.4 Demands of Future Production Management Systems

2.4.1 Target Management

Overview
Target management in combination with continual improvement processes (CIPs) is a basic tool of modern production management. Targets for turnover, yearly reduction of production costs, increasing productivity, or smaller batch sizes are set at the company management level. These targets generally are relevant for production; that is, they must be implemented and monitored at the production management level and therefore fall into the scope of the MES (Fig. 2.3).

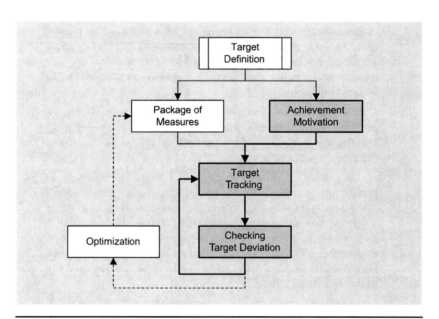

Figure 2.3 Target management—general workflow with emphasis on MES-relevant sections.

A modern MES must support target management as per Fig. 2.3, or even should constitute the core target management tool. Approaches for this can be found especially in the areas *motivation systems, target tracking,* and *analysis of target deviations.*

Motivation Systems Controlled by the MES
For the motivation of employees that is necessary to achieve a target, an MES can be introduced at hardly any extra expense. The *monitoring/ visualization* component in particular constitutes a suitable vehicle. The target set is, for example, "improvement of machine breakdown documentation." Displaying the existing degree of documentation on the MES terminal of the machine can prompt workers to input the reason for the breakdown. In addition, the degree of documentation (ratio of non-documented machine breakdowns to all machine breakdowns) can be visualized per machine or section via a group display. The group dynamic this creates can have a positive influence on all parties involved. Small incentives for active cooperation can include links to weather forecast or a daily updated cartoon on the MES terminal.

It is important that all displays and user interfaces of the MES are extremely user-friendly. Although it is difficult to map complex and extensive functions in a pleasant manner, the time invested here will contribute to acceptance of the system. Hard-to-understand user concepts always have a demotivating effect.

MES as a Target-Tracking Instrument
The most important aspect of target management is setting measurable and therefore verifiable goals. Verbally formulated targets that are impossible to verify contribute at best to a general good feeling but are hardly a suitable management tool. The measurement and visualization of all data arising in the production process are one of the tasks of the MES. Condensed data offered by the MES as *key performance indicators* (KPIs) constitute an especially useful management tool. The goal of the company's performance is, for example, to reduce the rework quota for a product to x percent and at the same time to increase the quantity to y percent. This goal can be pursued through suitable performance figures that are provided by the MES per production shift. Here, the timely publishing of the results is crucial. Deviations should not become known only days later but ideally should be published online or at least per shift for the production employees.

Information distribution will be looked at in more detail in Sec. 2.4.4, as an independent domain of the MES.

Analysis of Target Deviation in the MES
When the aforementioned KPIs demonstrate a deviation from the targets set, it is essential to uncover the reasons for this. The MES must offer suitable reports for this purpose that allow for simple analysis of the data. Many MES providers extend their systems to include

powerful reporting tools, or *business intelligence systems*. However, these systems are not really suitable to accompany short-term actions (which are actually the norm), such as the aforementioned reduction of reworking together with simultaneous increases in quantity. Their complexity makes it difficult, if not impossible, for employees to create reports in a timely manner, which means that the action is usually complete before the (supposedly accompanying) reporting tool has been adjusted.

The MES therefore must allow for analysis of the existing data with in-house tools and must include an easy-to-use toolset for the flexible adjustment of exceptional situations. Production workers must be enabled to administer the reporting that is relevant for them—only thus can a short-term analysis of target deviations be achieved.

The result of this analysis is a *continual improvement process* (CIP), and the target management control circuit is closed.

2.4.2 Integration of Applications and Data

Overview
The deficits of the nonintegrated systems described in Sec. 2.3.1 must be corrected using *integrated systems*. The *integration* concept refers to three fundamental topics:

- Applications
- Product data
- Production data

Application Integration
Many production workflows are today accompanied and supported by IT systems. However, as a rule, there are numerous systems created for special tasks and aimed at different user groups. This means that information is inevitably lost or existing information is sent to the incorrect addresses at the wrong time.

One concept for the necessary integration of applications is *collaborative production management* (CPM). As illustrated in Fig. 2.4, all user groups in the company, from the manager to the worker and from purchasing to marketing, should have access to the relevant data through an integrated software solution.

A technological approach to linking various applications is the concept of *service-oriented architecture* (SOA; see Sec. 7.1.6). By means of this architecture, a situation is achieved where *a process* and the related data can be mapped only *once in the company's IT*, and the software functions are made available to *all users* in their specific contexts. For example, customer orders are booked in the CRM of the sales department and *simultaneously* made available via a service of the ERP system or the MES.

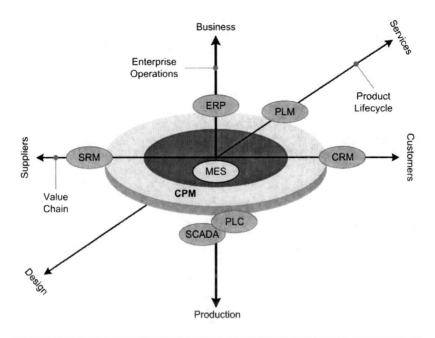

FIGURE 2.4 Scope of CPM in the context of adjacent systems (see also [ARC 2003]).

Product Data Integration
In accordance with the ISA level model (see also Sec. 3.2.1), the development process (i.e., design and construction of the product) and the production process are ranked at level 3, or the production management level. The ISA currently restricts its guidelines to the production process. In the sense of application integration (see also CPM in the preceding section), however, a link between the development process and production is desirable.

The development process up to planning of the process and ongoing change management (in the sense of a requirement process) are functions of the PLM system (see Sec. 3.5). The entire production (both the production of samples/prototypes and the subsequent production of products) is managed by the MES. Parts of change management (the requirements that are passed on to the PLM system are first defined in production and partially tested) are also rooted in the MES. Management of the *product definition data* (product definition management) is the link between both processes. This makes an MES a true *product management system* (PMS), which assumes *product definition management* as well as production (Fig. 2.5).

A product is defined by various documents and data sets. Some examples are work plans, parts lists, recipes, and process instructions. The first version of these data, a *blueprint* of the product, as it were, arises in the course of the development process. However, these data

Chapter Two

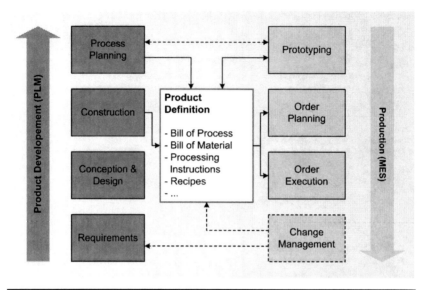

FIGURE 2.5 Product definition management as part of the MES.

are already required by the actual production machines and facilities for the production of a prototype and therefore already should be managed and maintained by the MES at this time.

Throughout the lifecycle of the product, changes to the original product definition may arise. Qualified change management (definition and testing of new products and definition of requirements for the PLM) as part of the MES must document these changes entirely and must, in turn, make them available for the development process. This also improves the agility of the entire company. The data of existing products or components of these products are made available to development as the basis for new developments. The cycle from product planning to delivery of a new product to the customer can be shortened considerably.

Going hand in hand with this, integration and harmonization of various data structures are necessary. In today's system landscapes, every individual application has an independent database. As a result, both redundancies and the risk of inconsistency arise.

The MES requires data from the adjacent systems for its functions. Much of this *master data* is expanded and made more detailed by the MES. This leads to the demand for the MES to have access to an exhaustive and consistent data model. Thus the MES is the natural database of the entire production and provides all other IT systems with the necessary master data. This ideal perception cannot be applied without untenable expense, especially in historically developed heterogeneous system landscapes. Here, *master data harmonization systems,* also known as *master data management* (MDM) *systems,* can provide a remedy.

Master Data Management (MDM)
Master data (e.g., customer and supplier address data and article data) are the basis for all IT processes and commercial records in the company. The quality and availability of these data for all IT systems accordingly are important. Especially for decentralized company structures (e.g., parent companies with independent domestic and foreign subsidiaries), harmonization and maintenance of the master data represent a true challenge. In such a structure, management of master data must be organized very flexibly. This means that systems and maintenance can be assumed both centrally and also by associated entities such as subsidiaries. Global master data objects and attributes are defined. If necessary, these data objects are extended by means of specific process-relevant attributes and saved centrally. After collection or maintenance, these global data are forwarded to the associated entities by means of a distribution mechanism. Additional attributes may be present in their systems, if needed, that are used exclusively on a local basis. (See [ZIMMERMANN 2005].)

Production Data Integration
The data arising in the production process are collected and archived by the MES. These data should be usable by various groups of people within and outside the company as quickly as possible after they are created:

- *Production status data.* The status of the current orders with regard to quantities, dates, and quality is made available to those responsible for production and to sales and marketing (or directly to the customer).
- *Operative order planning data.* For example, material requirements can be passed on directly to the suppliers with the desired delivery date. Sales and marketing (or the customer directly) receive information about planned delivery dates.
- *Facilities/machine status data.* Recording the status of facilities and machines is also an MES function. Based on these data, short-term disruptions can be remedied and processes for preventive servicing can be controlled by maintenance.

2.4.3 Real-Time Data Management

Introduction
The topic "shortening response times" can be found in almost every publication about production. This attitude of making all products (as well as services) available within an increasingly short time is in keeping with the spirit of the modern achievement-oriented society.

This way of thinking has far-reaching effects on production management: next to quality, adherence to delivery dates is given top priority. Urgent orders must be processed "in between jobs." Decisions about alternative production strategies or emergency concepts must be made quickly and substantiated on the basis of available information. The entire supply chain for material and preproducts must, of course, also provide the required flexibility so that production is not held up. Last but not least, all necessary systems and machines must be available 24 hours a day (and often on Saturdays or Sundays)—this requires effective management of disruptions and resilient data for preventive, anticipatory, or usage-dependent maintenance.

The concept of *real time* has various understandings depending on the respective domains. For example, while processing times of microseconds or milliseconds are expected of real-time systems in production technology, they may be in minutes in process engineering (see Sec. 3.4.3). Real-time data management in production must enable response times measurable in the range of seconds (e.g., for alarm systems and visualization of the current system status) to about 1 hour (e.g., statements about possible delivery times to a customer).

Event Management

Both events from the production process (e.g., messages coming directly from the systems/machines of production) and events that are generated in the MES itself (e.g., deviations between target and actual values; see "Early Warning Systems") must be processed immediately. Events need to be assessed and must be transmitted to the responsible decision makers in the context of "escalation management" (see "Escalation Management").

Early Warning Systems

Since order planning still can provide only a *target framework* for production parameters (largely dates, quantities, costs, and qualities), the MES must constantly compare the actual values that have arisen in the context of order processing with the targets and must generate warning messages (i.e., an event; see "Event Management") in case of deviations. It must be possible to parameterize the value of the tolerable deviation flexibly here. Associated with this are automatically transmitted details on the causes of the deviation. In the future, methods from multivariate statistics will be employed increasingly.

Multivariate Statistics

Multivariate statistics provides methods and processes that aid the preparation, tabular and graphic representation, and evaluation of complex data situations. Multivariate statistics processes (as compared with conventional statistics) are characterized by the fact that they allow

> the collective simultaneous analysis of numerous features and their expressions. Thus, when the expression of numerous features is observed in objects, all observed data can be evaluated jointly with the aid of multivariate statistics. The advantage as compared with individual, univariate analysis for each value is that the dependencies between the values observed can be taken into consideration. (See [ELPELT HARTUNG 2007].)

In an ideal situation, these deviations can be recognized in good time by calculating a *trend value*. Thus the MES anticipates a violation before it actually occurs and therefore provides the information to the responsible parties in order to ensure that suitable countermeasures can be taken in time. A simple example for such a trend calculation is the prediction of the production quantity at the end of a shift: if the cycle time is known, the quantity at the end of the shift in a discrete production process can be calculated continuously. This information can be made available both to the workers (e.g., visualization of the trend in a large display) and to those responsible for production. Management thus can step in early to take countermeasures (e.g., using additional staff resources and shortening the cycle time).

Cost-Control Management

The monitoring of costs occupies a special place in the monitoring and early warning function (see "Early Warning Systems"). The cost-effectiveness of the production process—for all workflows—is a decisive competitive factor. Not only the direct costs, such as material usage and time needed by personnel deployed, must be monitored, but also the overall costs for the article produced per workflow. As a result, *activity-based costing* methods will be used increasingly in the future to allow the cost of the product in the production process to be calculated as precisely as possible. Conventional cost-control systems do not take this into account. As a result, an incorrect impression of the actual profitability of orders and products may arise.

> **Activity-Based Costing**
> In this approach to costing, it is assumed that the resources of the company are used for providing services (i.e., for production in the sense of an MES). The costs of the resources are allocated to the activities that use these resources. The activity costs form the sum of the resource costs for an activity. The costs of the product ensue finally from the sum of the activity costs. Here, we refer to all cost components that cannot be calculated as individual costs, especially factory overheads. Production therefore is expressly included in the application of activity-based costing (unlike various other approaches to costing). (See [KRUMP 2003, p. 18ff].)

Escalation Management

Escalation management is closely linked to event management. On the basis of an escalation plan, an event is evaluated in the short term, and the escalation levels are activated in accordance with an existing hierarchy. Escalation plans are laid out in such a way that critical events in the framework of a defined workflow can be brought to a solution soon.

2.4.4 Information Management

Overview

The data collected and created by the MES (see Secs. 2.4.2 and 2.4.3) are of use only if the information gleaned from them is directed to the correct places. A major element in the extraction of information is compression. Information for the management level of a company must be greatly compressed in order to be absorbed at all. In extreme cases, thousands of reports or individual pieces of data are condensed into one index that describes the efficiency of the production department (or a sector of production).

Only structured data management makes it possible to use real-time information in making rapid decisions. In addition, the data must be available for correlation checking and more long-term analysis.

Key Figures

Significant key figures or KPIs are a tried and tested way of reducing the flood of information to a manageable amount. In the systems used today, key figures generally are produced at the end of a particular period of time in production (e.g., at the end of a shift or at the end of a day). The responsible parties never know until after the fact how well or badly their production has gone. Real-time control systems provide these key figures immediately, creating *online KPIs* (see also Sec. 2.4.3) by means of which measures can be taken earlier and therefore more effectively.

MES generally provide a set of *standard KPIs* such as availability, performance, quality rate, and *overall equipment efficiency* (OEE). There is a true conflict of beliefs in many companies about the significance of the key figures and especially about the issue of which format which key figures should be provided in. However, there are no straightforward answers for these questions. Whether one or more key figures are needed to inform management or what formula should be used to determine an OEE, for example, depends largely on the type of production (e.g., discrete assembly or continual process, high or low degree of automation, etc.) and also on which control factors management sees as actually important (i.e., which key is the right one to use for assessing production). This situation leads to the following demand on an MES: in addition to a set of standardized key figures, the MES also must be able to calculate project-specific

KPIs at the user's request. The guidelines for calculating these KPIs should be easy to modify, preferably by the user himself or herself. In order to provide continuity, the system must be capable of displaying existing and previously used key figures for control processes.

Correlation Systems
Until now, the interdependency of data has been treated somewhat negligently. Multivariate statistics (see also Sec. 2.4.3) will be drawn on with increasing frequency in the future for this purpose. This will reduce the flood of data to the essentials, linkages will be recognized, and on this basis, long-term decisions can be made.

Distribution of Information
With an integrated system, it is possible to display the relevant information in a central "cockpit." However, this type of distribution forces the users to refer to the system. The users are obligated to refer to the system regularly and actively. Important information additionally should be transmitted by the MES actively and specifically (*proactive communication* by the MES). Compression and transfer of information ideally are carried out by divisions; for example, production management receives all compressed information for that division, general management and controlling receive only the cost deviations, and sales and marketing receive deviations in date.

Web technologies, especially the medium of e-mail, have established themselves as a vehicle for the distribution of this information. This trend will become increasingly pronounced because of the given advantages (e.g., high speed, asynchronous communication, and display on mobile devices). The MES therefore must be capable of distributing production data as flexibly as possible to various recipients. In order to avoid an undesirable flood of information (i.e., a message from the MES should not be perceived as an annoyance), the recipient of the information generally should be in control of distribution. The marketing employee should determine, for example, at what time of the day he or she would like to read a report about current order planning and how this report should be presented. For automatic communication between IT systems (e.g., blanket purchase orders for a raw material), standardization of the data structures will continue to develop. Here, the MES must be able to support different data structures and formats. This can be done, for example, using templates for messages.

2.4.5 Compliance Management

Compliance generally means the observance of all relevant laws as well as internal and external guidelines in companies. Compliance is a central factor for the success of a company today. Numerous current examples show that noncompliance with normative basic conditions is associated with high liability risks and enormous damage to the image of a company. The slogan "Comply or die" may sound blunt, but it clearly shows how relevant this topic is.

For which laws and guidelines must the company's management produce the necessary compliance? The number and content of guidelines change constantly and at an ever-increasing pace; therefore, there is no blanket answer to this question. The following grouping of the provisions should facilitate an overview:

- *Ethical and conduct guidelines.* These include, for example, guidelines for the company's management on its conduct with regard to customers and suppliers and on communication (e.g., code of conduct for top management), which are generally drawn up in the form of company-specific "recommendations."
- *Financial and accounting guidelines.* The best-known provisions in this area are Basel II in Europe and the Sarbanes-Oxley Act (SOX) in the United States.
- *Quality assurance guidelines.* These guidelines pertain to securing the highest possible level of quality for a product and therefore apply in the sphere of the MES. Especially for pharmaceutical products and foodstuffs, it must be ensured that the customer is not endangered. Examples for such guidelines are the DIN EN ISO 9001:2000 standards, the Food and Drug Administration (FDA) security standards (e.g., the often-quoted Guideline 21, CFR Part 11; see Sec. 3.2.4), and the EU directives (e.g., as per EU Directive 178/2002, all production and shipping stages for foodstuffs and feed must be absolutely traceable).
- *Environmental protection guidelines.* This area is gaining increasing importance, especially through the ongoing discussions about CO_2 emissions.
- *Safety guidelines.* This group includes guidelines on occupational health and safety, which differ greatly depending on the country and the company, as well as data protection guidelines (Fig. 2.6).

A portion of these guidelines (see "Quality assurance guidelines") falls quite clearly into the scope of the MES; that is, the MES can verify and document compliance with these guidelines. Thus three possible solutions arise from the IT point of view:

1. The MES takes on only the enforcement, monitoring, and documentation of product quality guidelines, that is, those guidelines that directly affect production. All other guidelines are processed with the aid of other tools and methods.
2. The enforcement, monitoring, and documentation of all guidelines within the company are managed by an independent

Factory of the Future   27

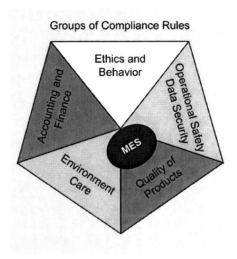

FIGURE 2.6 Groups of compliance rules and scope of the MES.

compliance tool. In this case, the MES must provide the compliance tool with the relevant information from the production process.

3. The MES assumes the enforcement, monitoring, and documentation of all guidelines within the company.

2.4.6 Lean Sigma and MES

Introduction

The buzzword *lean manufacturing* goes back to the measures formulated by Toyota in the 1950s to increase efficiency in production. At that time, there was practically no such thing as IT, and the measures were aimed mainly at organizational and training aspects. This concept, known as the *Toyota production system* (TPS) consists mainly of

- *Avoiding or reducing sources of loss.* Reducing waiting, storage, and transport times, rational workspace design, merging of production steps, and avoiding indirect activities (not contributing to value creation).
- *Synchronization of processes.* The production steps of the order processes are aligned with one another using demand-oriented planning (pull principle) and are synchronized.
- *Standardization of processes.* Process flows are modeled on fixed guidelines. Adherence is subject to regular audits.

- *Preventing failures.* Eliminating or reducing rejects and reworks.
- *Improvement of machine productivity.* Through preventive and usage-dependent maintenance, analysis of weaknesses and resulting CIPs, etc.
- *Continual training and further education of employees* (Fig. 2.7).

Lean Sigma (or *lean 6Sigma*) is a new term for concepts on increasing efficiency in production, that is, a way to produce products increasingly quickly, optimally, and with as few defects as possible. The combination of *lean manufacturing* (on the basis of the TPS) and the statistical methods of the *6Sigma* are consolidated under the term *lean Sigma*. The methods of lean Sigma should contribute to the faster, more precise, and less expensive production and implementation of products and processes. Many consultants offer their services in this regard (e.g., organizational measures, training, etc.).

Lean Sigma Projects

The TPS concept was enriched in the 1980s by a Motorola initiative with the DMAIC method (DMAIC stands for *d*efine, *m*easure, *a*nalyze, *i*mprove, and *c*ontrol and defines a control cycle for the implementation

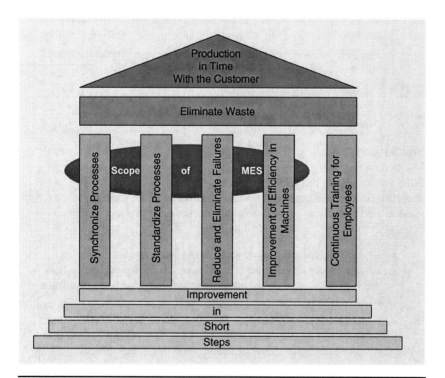

FIGURE 2.7 Concept and goal of the Toyota production system with the scope of an MES.

of measures) for zero-defect production. A lean 6Sigma project is also controlled using this DMAIC method. In addition, statistical methods and their associated instruments (SPC/SQC) are of central importance here. Consultants generally use specialists who train employees in the use of the tools mentioned. Depending on the level of training, the employee receives a "degree" (master black belt, black belt, green belt, or yellow belt). The goal of a lean 6Sigma project generally is to improve a key figure, for example, reducing the failure quota. The zero-defect production concept is so important because defect costs within a company often reach a considerable level. Therefore, the quality level in a company must closely approach the 6Sigma level.

The MES as a Vehicle for Lean Sigma
Astonishingly, MES advocates and providers were late in recognizing that MES provides the main tools for the implementation of these concepts. It is absolutely possible to affirm that a functioning MES is a condition for achieving goals and implementing measures from the lean Sigma concept. An operative planning system, the functional core of an MES, reduces waiting, storage, and transport times by synchronizing production processes. MES also ensures standardized processes—employees are guided with electronic information, which contributes significantly to improving productivity.

What is missing for the breakthrough of this knowledge on the broader market is close cooperation between consultants and production management system providers. Synergy effects would result from such a symbiosis, which would be advantageous for all parties involved. The factors that have prevented such close cooperation are as follows:

- Consultants generally do not have sufficient knowledge about functions and workflows of an MES. They are not familiar with using the tools. They are generally restricted to organizational services and employee training.
- MES providers generally are concerned with the problems of software development and the technological basis of the system. Therefore, the products lack systematic orientation toward the conditions required to implement lean Sigma.

Both sides must increase the quality of their services. Consultants should be able to identify with MES systems and acquire the necessary know-how to do so. On the side of the product provider, a universal consistent design is lacking. It is very unlikely that existing software components can be embedded successfully into such a design. Consequently, the products need to be developed from scratch in order to achieve the necessary standard. The factory of the future requires qualified integrated production management systems that must be accompanied by a consultant who is just as qualified in order to achieve its full potential.

2.5 Summary

The MES is developing into a strategic instrument for flexible, networked production. All production management tasks are summarized in an *integrated platform*. The MES therefore is not a loose collection of software components (patchwork) but rather an integrated system that allows the modular use of individual functions and makes these functions available to other software systems in the company, for example, by means of a service-oriented architecture.

As a database, the MES requires a *complete and consistent data model* that contains both a map of production with all its resources and the *product data* (or rather, the data for product definition). Therefore, the MES must be closely integrated to the PLM system and work hand in hand with it. The master data contained in the MES, for example, on the articles to be produced, are managed through *master data management* and are also made available to other IT systems.

The maxim "Faster, better, cheaper" also poses new challenges for production management. In part, the demand for ever-shorter cycle times and higher flexibility can be fulfilled only by transferring tasks from the company management level (level 4 on the ISA model) to the production management level (level 3 on the ISA model). The MES therefore must be developed further to form a *product management system* (PMS). Thus, on top of the existing core tasks (i.e., management of production resources and fine planning/control of production orders), it will take on some additional tasks:

- *Complete planning function,* including material and resource planning
- *Real-time data management* with early warning systems and cost control (keyword—activity-based costing) as the basis for fast decisions
- *Information management* on the basis of flexible key figures (KPIs) and correlation systems (keyword—multivariate statistics) and distribution of information also through proactive data transmission
- *Compliance management* as a response to the fast-growing number of guidelines and regulations

Through the functions described, the MES becomes a central strategic tool for implementing the requirements of the factory of the future.

CHAPTER 3
Concepts and Technologies

3.1 Commonalities between Existing Approaches and MES

Before an approach to solving the demands for the factory of the future identified in Chap. 2 can be developed, the merits of existing standards and technologies that could be called on to solve the problem must be examined. Following an analysis of the existing norms and guidelines, recommendations for a manufacturing execution system (MES) are described in more detail. Subsequently, information systems from neighboring sectors are introduced. Then a comparison is made with the MES. Finally, product lifecycle management (PLM) is examined. Commonalities between both approaches are indicated.

3.2 Norms and Guidelines

3.2.1 ISA

The ISA was founded in 1945 in Pittsburgh. Originally, the acronym stood for *Instrument Society of America*. The 18 founding member companies from the process technology sector aimed to be better able to represent and implement their common interests.

Today, the organization's acronym stands for the *Instrumentation, Systems, and Automation Society*. The ISA is active internationally and currently has more than 28,000 members from more than 100 countries. Its tasks and goals include organizing conferences and trade fairs and drawing up guidelines for industrial process instrumentation. Many of the writings of other organizations are based on ISA concepts.

Among others, two guidelines were passed by the ISA that are relevant or even essential for production-oriented systems such as an

MES. The following explanations examine the content of both guidelines in more detail.

ISA S88

The first part of this standard was passed in 1995 and was declared a national norm (equivalent to DIN in Germany) by the U.S. institute for the standardization of industrial process guidelines, the American National Standards Institute (ANSI), and therefore can be seen as the most important part (see also Sec. 3.2.2). In this part, reference models for batch control in the process industry were defined. The connections and relationships between the models and the processes are explained. The core consists of the recipe operation mode, which allows for a division between plant structuring and process structuring (Fig. 3.1).

A batch/lot process creates a defined quantity of a product, which is composed of one or more raw materials and is manufactured in one or more devices in a defined order. Production instructions are indicated in the (basic) recipe. The recipe consists of a set sequence of processes (operations)—defined in partial recipes and executable in parts of the facilities (e.g., reactor, dosing stations, mills, etc.)—that, in turn, consist of a set order of functions (phases)—defined as partial recipe steps (e.g., dosing, heating, mixing, etc.). Every step must be accompanied by a list of parameters. The type and extent of the parameters depend on the function called up by the step (e.g., ingredient with the proportionate amount, temperature control with target value, etc.).

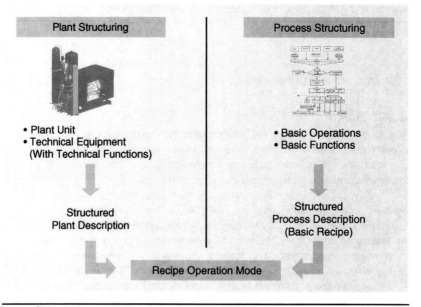

FIGURE 3.1 Recipe operation mode as per ISA S88.

Since the basic recipes are provided independently of the plant, the recipe header of the partial recipe must contain a list of facilities suitable for production.

The second part of this standard was passed in 2001. This part defines data models and their structures for batch control in the process industry. This should facilitate the standardization of communication within and between the individual batch controls.

The third part was released in 2003. Here, the focus is on the possibilities for standard data analysis and suitable presentation of the results.

The last and latest part dates from 2006. This part specifies the recording and archiving of data for process technology in detail. The aim is to use a reference model to facilitate the development of applications for data storage and/or data exchange. At the same time, this standard strives to ensure that the tools have at least these specified functions for data recovery, data analysis, reporting functions, etc.

ISA S95

In production, separate realms of information have developed over the years with different mind-sets and disciplines: the enterprise resource planning (ERP) systems of the economic sector, the MES in production, and the actual level of automation with sensors, actuators, controls, etc. However, companies today need a continuous flow of information throughout the entire operation. Although the business and management systems have different abilities and objectives, the production level requires, for example, production data in real time, whereas management thinks more in the medium or long term, but they still need the same information for planning production, evaluating production performance and capacity, and making projections for maintenance work.

ISA S95 defines terminology and models that are used for integrating ERP systems at a business level with automation systems at a production level. The standard, which like S88 was declared a national norm by the ANSI, is divided into the following parts:

> Part 1 was published in the year 2000 and contains the basic terminology and models for defining interfaces between business processes and the process and production management system.
>
> Part 2, dating from 2001, defines, together with the first part, the contents of the interface between control functions in production and company management.
>
> Part 3, published in 2005, provides detailed definitions of the main activities of production, maintenance, warehousing, and quality control (Fig. 3.2).

The functional division between ERP and MES level is of significance for the following explanations. These are described in the first part of S95 and are as follows:

34 Chapter Three

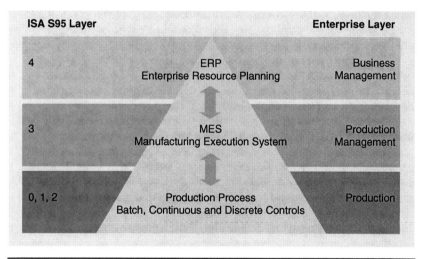

FIGURE 3.2 Level model according to ISA S95.

Level 4 tasks [ISA S95-1, p. 19ff]:

- Management and maintenance of raw materials and replacement parts. Provision of master data for the purchase of raw materials and replacement parts.
- Management and maintenance of energy resources.
- Management and maintenance of master data needed for preventive and foreseeable maintenance work and management and maintenance of personal master data for human resources department.
- Adopting a rough plan for production and revising the rough plan based on the resources available and scheduled maintenance work.
- Maintenance of warehouse master data.
- Determining optimal stocks, energy supplies, replacement part stocks, and stock in production. This also includes examining material availability with regard to order release for production [material resource planning (MRP) run].

Level 3 tasks [ISA S95-1, p. 20]:

- Evaluation of production-relevant data, including determining real production costs.
- Management and maintenance of data related to production, inventory, personnel, raw materials, replacement parts, and energy. Furthermore, the management and maintenance of

all additional personnel information/functions, such as timekeeping, holiday calendar, human resources planning, qualifications of employees, etc.

- Establishment and optimization of fine planning for every division. This also includes any possible maintenance, transport times, and all other production-relevant tasks.
- Reservation of resources relevant to orders (facilities, staff, material, etc.). Any changes (e.g., machine breakdown) should be recorded promptly so that plans can be altered, if necessary. Data must be archived. The production orders are transmitted to the resources available by the system. They are redistributed automatically in the case of any disruption.
- General process monitoring functions (alarm management, tracking, tracing, etc.).
- Provision of functions for quality management, compiling key figures, timekeeping, and maintenance management.

In summary, it can be said that the specification offers the reader much scope for interpretation. Many areas are described insufficiently or not at all. All unspecified points are allocated automatically to level 4. This also implies that product definition cannot be allocated clearly to a specific level.

3.2.2 IEC

The International Electrotechnical Commission (IEC) was founded in 1906 in London. Since 1948, its headquarters have been in Geneva. The IEC was involved largely with standardizing norms for units of measurement. It was also the first to recommend a system of standards that later became the International System of Units (SI units).

The IEC regulations incorporate all aspects of electrotechnology, including the generation and distribution of energy, electronics, magnetism and electromagnetism, electroacoustics, and multimedia and telecommunications. General disciplines such as terminology and symbols, electromagnetic compatibility, measurement technology and operating performance, reliability, design and development, and safety and environment are also included.

IEC 61512
Not only was ISA S88 (see Sec. 3.2.1) declared a national norm in the United States, but it also was declared a universal international norm by the IEC in 1999 [IEC 61512].

IEC 62264
Like ISA S88, ISA S95 not only is a national norm in the United States, but it also has been valid internationally since 2003 as an IEC guideline [IEC 62264].

3.2.3 VDI

The Association of German Engineers [Verein Deutscher Ingenieure (VDI)] is a scientific association founded in 1856 in Alexisbad. It has its headquarters in Düsseldorf now. Along with engineers from various specializations, its members increasingly include natural scientists and computer scientists.

The VDI thus is one of the largest technically oriented associations in the world. Since its founding, it has built up a body of technical rules and regulations that today comprises over 1,700 current guidelines covering the vast field of technology.

VDI 5600

The VDI has only recently become involved in the area of process technology and in particular MES. The VDI 5600 directive, the only directive relevant to MES, was drawn up in accordance with ISA and Manufacturing Enterprise Solutions Association (MESA) standards but has not yet been released.

The directive is intended to provide support to the potential MES user in all phases, from the selection of a suitable system to operating the facility, in a company-neutral and professionally founded manner. The tasks of an MES, from fine planning to staff management, are roughly described. In another part, the significance of MES and the support of various company processes are examined in more detail.

In summary, the document contains mainly reflections, explanations, and updates with regard to existing concepts.

3.2.4 FDA

The Food and Drug Administration (FDA) is a public agency of the U.S. Department of Health and Human Services. It is responsible for the issuance and observance of safety provisions and standards across the entire production chain of foodstuffs, pharmaceutical products, etc. It monitors the production, import, transport, storage, and sale of such products. Implementation of the standards should ensure that the American consumer is protected, regardless of whether the respective product is produced in the United States or not. Only the target market (United States) is significant. Accordingly, compliance with these conditions is essential worldwide for all manufacturers who wish to supply their products in the United States. The agency reserves the right to carry out audits worldwide.

However, the regulations of the FDA are of interest (or essential) not only for the originally relevant sectors. Some approaches are absolutely suitable for application to other sectors, such as manufacturing technology.

FDA 21 CFR Part 11

FDA 21 CFR Part 11 (Electronic records; electronic signatures) defines criteria in accordance with which the FDA will accept the use of electronic records and signatures as the equivalent of paper data records and signatures. The regulation was the result of six years of work (including initiatives from the industry). The goal was to establish criteria for all companies under supervision of the FDA regarding how paperless electronic recording systems could be run in line with good manufacturing practices. These include

- GMP 21 CFR 110 (Good manufacturing practices for foodstuffs)
- GMP 21 CFR 210 (Good manufacturing practices for manufacturing, subsequent processing, packing, or holding of drugs)
- GMP 21 CFR 211 (Good manufacturing practices for finished pharmaceuticals)
- GMP 21 CFR 820 (Good manufacturing practices for medical devices)

The standards are considered legally binding in the United States. The general criteria drawn up there regarding a quality management system and process are largely in agreement with DIN EN ISO 9000ff (series of standards regarding quality management). The processes described are also valid for an MES.

3.2.5 NAMUR

General Information on NAMUR

NAMUR (International User Association of Automation Technology in Process Industries) was founded in Leverkusen, Germany, in 1949. It is an association of user firms in the process management technology sector. Manufacturers of process management systems, hardware, or software are not eligible to become members. The NAMUR recommendations [Namur Empfehlung (NE) = Namur Recommendation; Namur Arbeitsblatt (NA) = Namur Process Sheet] include practice reports and work documents produced by NAMUR for its members from the circle of users for facultative use. These papers are not to be seen as norms or directives but rather as complements to these.

NE 33

The NE 33 recommendation (Requirements to be met by systems for recipe-based operations) is intended to unify the terms and concepts used in the process industry and the supplier industries with regard to recipe-based operations based on ISA S88 (see Sec. 3.2.1). The recommendation places special emphasis on structural aspects. On the other hand, little or nothing is said about the realization of functions.

The aim is to present universal instructions on recipe-based operations. The demands on systems using recipe-based operations are formulated based on a universal concept for operating process technology systems. Not only is the concept valid for discretely operating systems, but it also can be transferred to continually operating systems, especially with regard to startup procedures and rundowns, working points, modifications, etc.

In NE 33, the running of continually operating systems is not dealt with in more detail. The concept refers to normal usage, as well as to the mastery of unusual circumstances. The methods recommended here are applicable irrespective of the degree of automation and include the division of labor between humans and automation systems.

NA 94
In the scope of the implementation of supply-chain management (SCM) systems, the technical and logistic production processes of plants are optimized. The new processes will raise the standards for production-oriented data-processing systems with the superimposed enterprise resource planning (ERP) systems.

Based on ISA S95 Part 1, a data model is developed in recommendation NA 94 [Machine Data Acquisition (MDA): Functions and sample solutions at company management level] in which the functions of an MES and the information flows to be covered are described.

3.3 Recommendations

3.3.1 MESA
The Manufacturing Enterprise Solutions Association (MESA) is a U.S. industrial association with a focus on improving business processes in the production sector through optimization of existing applications and introduction of innovative information systems. Here, both the vertical and horizontal integration of information systems plays an important part. Just a short time before the ISA (see Sec. 3.2.1), MESA was the first organization to devote itself to the topic of MES.

MESA Guidelines
MESA has a range of information on the topics of production management, product creation, quality management, and production optimization for manufacturing plants and solution providers. Here, the integration of production-oriented systems (execution processes) is of particular significance.

The 11 function groups of an MES are as follows, according to MESA:

1. *Fine planning of workflow.* This group envisions optimal sequence planning with regard to the relevant basic conditions (setup times, processing time, etc.) based on the resources available.
2. *Resource management with status maintenance.* Management and monitoring of the relevant resources (staff, machines, tools, etc.).
3. *Production unit control.* Control of the flow of production units based on orders, batches, etc. Events during ongoing produ-ction are responded to immediately, and if necessary, the plan is adjusted.
4. *Information control.* All information relevant to the production process (CAD, designs, test specifications, environmental compliance requirements, safety instructions, etc.) is made accessible to the staff at the right time and right place. Staff can use the system to record deviations.
5. *Operating data logging.* Automatic or manual logging of all production-related operating data linked with the production unit.
6. *Staff management.* Recording of staff working hours and potential to edit in case of absence, holiday, etc.
7. *Quality management.* Analyses of production-related measurement data in real time in order to safeguard product quality and be able to identify problems and weak points in good time.
8. *Process management.* Monitoring of the actual production process, including alarm management functions.
9. *Maintenance management.* Recording the use of operating material and hours of use in order to initiate periodic and preventive maintenance tasks. The system also supports the execution of maintenance.
10. *Lot traceability.* Recording of all production-related data across the entire production chain to ensure that every product manufactured is traceable.
11. *Performance analysis.* From the manufactured sizes to down time, disruptions, piece counters, etc., managerial key figures are produced promptly, in real time, if feasible, in order to allow for simple assessment of production efficiency, detection of problems, etc. Display in various diagram formats is made available to the user.

3.3.2 VDA

The German Association of the Automotive Industry [Verband der Automobilindustrie (VDA)], founded in 1901 in Eisenach, consists of car manufacturers and their development partners, the suppliers.

The umbrella organization also includes manufacturers of trailers, superstructures, and containers. The joint organization for car manufacturers and suppliers is unique in the world.

The VDA promotes the interests of the entire sector nationally and internationally in all areas of the automotive sector, for example, in economic, transport, and environmental policies; technical legislation; standardization; and quality assurance. The VDA, too, has recently recognized the importance of MES for the entire sector and has responded accordingly with a body of regulations. The document [VDA 2008] is aimed at automobile suppliers with a focus on quality assurance.

3.3.3 VDMA

In 1890, the foundations were laid for the subsequent German Engineering Federation [Verband Maschinen- und Anlagenbau (VDMA)] with the founding of the Association of Engineering Institutes of Rhineland-Westphalia (Verein Rheinisch-Westfälischer Maschinenbauanstalten) in Düsseldorf. The VDMA represents mainly medium-sized companies in the industrial goods industry and is currently the largest industrial association in Germany. The entire process chain is mapped out in the association—from components to facilities, from system providers and system integrators to service providers. The association is concerned with research and technical regulations and environmental, technological, and energy-related topics [VDMA 2008].

The VDMA has been examining MES since 2004. Current efforts are focused on promoting the establishment of MES as an international standard. Here, it is intended to participate in the relevant standardizing bodies in order to best represent the interests of the industry. This approach should take into account the various aspects, from the establishment of standards to the different application and implementation scenarios in industry (e.g., engineering, process industry, chemical industry, processing sectors, etc.).

3.3.4 ZVEI

The German Electrical and Electronic Manufacturers' Association [Zentralverband Elektrotechnik- und Elektronikindustrie e. V. (ZVEI)], founded in 1918, the association of the economic branch of the electronics industry in Germany, is based in Frankfurt am Main. It represents political and technological interests at a national and international level and supports international standardization projects.

After the VDMA, the ZVEI is the second largest industrial association in Germany. It provides its members with targeted information about the general economic, technical, and legal conditions for the electronics industry in Germany. Through study groups and in cooperation with other associations and societies, it has a direct impact on the drawing up of guidelines, formation of norms, etc.

This association also has recently recognized the significance of MES and has responded accordingly. In addition to intentional participation in standardizing bodies, the association organizes seminars and podium discussions on MES.

3.4 Adjacent Areas

3.4.1 Historical Development of ERP/PPS Systems

The ERP and production planning and scheduling (PPS) systems on the market at the moment are used largely for administration and accounting with the core functions of financial management and material management. Looking back over the past few decades, it can be seen that various electronic systems have been developed to make industrial production processes more effective. IBM made a start in the 1970s with accounting systems on mainframe computers. However, not every idea and development caught on. Approaches such as computer-integrated manufacturing (CIM) were unable to establish themselves. In the past, this was partly because the available computing power of the hardware was not nearly sufficient to implement the approaches appropriately on the software front. The ideas were ahead of their time. Today, sufficiently high-performing systems are available at moderate prices so that complex, computationally intense processes can be realized in an economically sensible way.

As well as these approaches from the management and administration levels, automation technology also has developed in the form of compact controls, computerized numerical control (CNC) machines, and use and monitoring components. Each of these levels has special tasks and criteria to fulfill. Cooperation and interaction are not easily possible.

MES closes this gap between the ERP level and actual production and thus not only acts as a link but also provides a range of additional significant functions that cannot be performed by the systems of the other levels.

3.4.2 ERP/PPS Systems

In Sec. 3.2.1 we examined more closely the distribution of tasks between level 4 and level 3 as per [ISA S95-1]. The tasks of an ERP system can be derived directly from this.

The tasks of ERP and PPS systems are not so easily differentiated. The approach of a PPS system was originally to support the user in production planning and control and to assume the data administration associated with this. Based on the definition, the aim of PPS systems really should be the realization of shorter processing times, adherence to dates, optimal stock levels, and the economic usage of resources. On the other hand, ERP systems provide modules for accounting, such as payroll accounting.

However, the reality looks different. Transitions between the systems are fluid. Many ERP systems today provide functions for production planning. On the other hand, many PPS system manufacturers are adding classic ERP functions such as accounting or purchasing to their software.

The tasks of a PPS system look very similar to those of an MES at first glance. However, ERP and PPS systems provide production employees with relevant information to only a limited extent. In addition, this information is often not provided promptly. Regulating processes in real time with regard to target values is therefore not possible.

3.4.3 Process Management Systems

Process management systems generally are used to run technical systems. The term was coined in the 1970s when the first proprietary systems for managing refineries and crackers came into being. These systems were both hardware- and software-encapsulated and were tailored to the special demands of the respective applications. Data exchange occurred exclusively within the system.

Until then, the NAMUR level model (Fig. 3.3) was the organizational plan for process management technology. The process

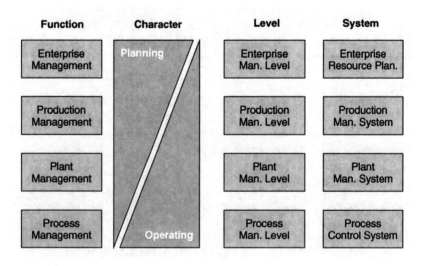

FIGURE 3.3 Hierarchical level model from NAMUR [NE 59].

management system therefore finds itself at the lowest level of the process management level. Communication between the levels is increasingly gaining importance. In this area, the systems also grow closer and closer together. Transitions are fluid. On the production level, an MES should be implemented as a management system.

A process management system is distinguished by the following core characteristics:

- *Real-time ability.* The real-time ability of a system is composed of the punctuality and simultaneousness of the system. *Punctuality* means the capability of the process computer to react within a defined period of time in the course of a technical process [DIN 44300]. The result here is that the response time of the automation system must be shorter than the actual process time. *Simultaneousness* means the quasi-parallel processing of numerous computing processes. For example, the regular intervals for an industrial titration process could be measured in seconds, whereas the intervals for a fighter plane would have to be measurable in microseconds. However, both systems are real-time-enabled with respect to the special area of application.

- *High availability through redundancy.* For complex industrial processes that are automated by process management systems, disruptions or short-term interruptions of workflows cause high failure costs, in the worst case, even danger to people or the environment. In order to increase availability, at least the critical parts of the system must be designed with redundancy. This means that the functions (e.g., field bus, sensor, PLC, etc.) should be taken over by reserve components without interruption in the case of a malfunction.

- *Openness and interoperability. Openness* means that the external interfaces and system characteristics of the respective manufacturer are disclosed so that it is possible to link any applications from other manufacturers. *Interoperability* guarantees that different components (e.g., field devices) from different manufacturers can be used together without additional outlay. Both characteristics are the basic conditions for vertical communication in the company.

- *Universality of an entire system.* Process information that is known at any point in the system must be accessible to every other component at any time without additional outlay.

Principal Structure of a Process Management System

The typical structure of today's decentralized process management systems is shown in Fig. 3.4:

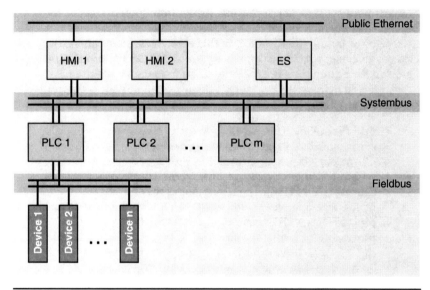

FIGURE 3.4 Principal structure of a process management system.

- *Process-oriented component.* A process-oriented component is a process computer, usually set up as a programmable logic controller (PLC), on which the functions having a direct effect on the process (e.g., PID regulator) are run. Furthermore, all sensors and actuators are connected to it. This can be done either on a point-to-point network or on a field bus.

- *Display and user component.* The display and user component creates the interface between the management system and the person running the system. Another term for this is *human-machine interface* (HMI) or simply *process visualization*. As well as installation images, there are also views for recipe administration (see Sec. 3.2.1), alarm management, etc.

- *Engineering component* (EC). The EC or engineering station (ES) allows the system functionality to be configured. This includes configuration of process-based components as well as display and user components.

- *System bus.* The system bus connects process-oriented components with each other and with the other components of the process management system such as display and user components and EC. This should not be confused with the subordinate field bus for connecting sensors and actuators to the controls. The system bus is always executed redundantly.

- *Open operation bus or plant bus.* This bus is the open interface of the system to the superimposed systems. The de facto standard for this is Ethernet with the transmission control protocol/Internet protocol (TCP/IP). The execution of this communication medium is not redundant.

3.4.4 SCADA Systems

As already mentioned, the tasks and functions of the concepts and technologies presented are interlinked and/or fluid. The systems have emerged from a wide range of approaches and sectors and have grown over the years.

Known as *supervisory control and data acquisition* (SCADA), the concept has arisen for the monitoring and control of technical processes. These systems include an HMI for the display of process images and the inputting of target values, etc. (process visualization). In addition, these systems also feature extensive alarm management and options for data archiving. Because of the areas of application (e.g., production technology, building control, etc.), these systems are seldom run redundantly, in comparison with process management systems.

Today, individual tasks are assumed to some extent by the MES, such as the administration of target values, which are transmitted order-specifically to the SCADA systems. MES take over compressed actual values from them for evaluation and archiving. As mentioned previously, the transfers are fluid and are also assumed by SCADA systems with expanded functions.

3.4.5 Simulation Systems

With modeling and simulation, a problem-solving process is transferred from reality to an abstract image and is solved using this image. A model of the reality of a system, of an object, etc. is needed as the basis for a simulation. Therefore, a model must be made in advance. If a new model is being developed, we speak of *modeling*. If an existing model is being adapted so that statements can be made about the problem that is to be solved, the parameters of the model need only to be set to either the actual situation or the desired target situation and varied as appropriate. The model and/or the simulation results then can be used to draw conclusions as to the problem and its solution (Fig. 3.5). Static evaluations then can follow, provided that stochastic processes have been simulated.

Simulation can provide insights on systems, for which real experiments are either impossible or would result in considerably higher expenses. The reasons for this are

- System behavior too slow or too fast (e.g., continental drift, nuclear reactions).
- System too large or too small (e.g., galaxies, atoms).
- A real system is unavailable (e.g., fusion process).
- A real system is nonexistent (e.g., product to be developed).

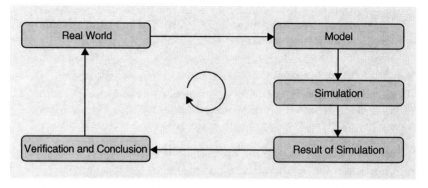

FIGURE 3.5 Closed circuit of model formation and simulation.

- A real system would be severely disrupted (e.g., stock market) or destroyed (e.g., crash test).
- A real system would be too expensive (e.g., aviation and space technology).
- The experiment is too dangerous (e.g., danger to humans and the ecosystem).

In an actual simulation, experiments are carried out on a model in order to gain knowledge about the real system, or the simulation results then are transferred to the real situation. In the context of simulation, we talk of the system that is to be simulated and of a simulator as the implementation or realization of a simulation model. This latter represents an abstraction of the system to be simulated.

Simulation is also becoming increasingly important in the area of automation technology. As we will see in Chaps. 4 and 6, it plays an important role in production planning and control. For example, there could be a considerable impact if an existing manufacturing plan is changed in the short term by a new order of higher priority. In such a case, it is helpful for those responsible for production if an MES is not only able to generate an optimal production plan but also can simulate scenarios and their effects. Thus the impact (e.g., what order is postponed, changes in cycle time, etc.) of introducing a priority order can be assessed in advance.

3.5 Product Lifecycle Management

3.5.1 Historical Development

In the past 15 years, information technology (IT) has moved not only into the production process but also into the product development process. Product development thus has changed dramatically in this period. Until 20 years ago, products were constructed and developed exclusively on the drawing board. The keystone of IT-supported development came about in the 1980s with the first two-dimensional (2D) computer-aided design (CAD) systems. However, these were merely an equivalent to the drawing board in terms of functionality and options. The working methodology for construction remained unchanged. The advantages of this new development in terms of changes, copying partial constructions, and manageability of drawings were obvious and brought about the rapid breakthrough of this innovation.

The real changes with regard to work methodology in construction came only after the availability of powerful computers and the associated possibility of three-dimensional (3D) designing using 3D CAD systems. Here, products are no longer designed in 2D in different drafts but rather are developed directly as 3D models. Thus they contain considerably more information than conventional drawings. For example, unwanted intersections between the gearwheels and the casing of a gear construction can be recognized in the development phase. Furthermore, 3D models offer numerous other options for simulation and manufacture.

While in the 1980s it was still customary to archive design drawings created with CAD on paper as before, with the advent of 3D CAD, a paradigm shift became necessary with regard to archiving methods. 3D models can be usefully managed only digitally. However, in order to still maintain an overview in these virtual archives, software approaches for the archiving and management of the models were developed in the 1990s. These product data management systems (PDMs) have since found universal application in product lifecycle management (PLM) and are therefore fixed components of respective systems.

Large quantities of data and documents arise from the idea to the development to the announcement of a product. The software tools necessary for the generation and administration of information have developed exponentially over the course of the past few years. PLM is a holistic concept for the IT-supported organization of all product and development data across the entire lifecycle.

3.5.2 Product Model

In order to realize the approaches of PLM, an integrated product model is essential. This contains the formal description of all information on a product across all phases of the product's lifecycle in one model [VDI 2219]. The result is that the PLM product model is a hub that must provide all relevant data on the product. However, in order to cover the information requirements of all departments and employees involved, these data must include not only technical but also organizational information. A model consists of various model components. These can map individual product sections or include various technologies such as pneumatics and hydraulics. The integrated product model merges these partial models into an overall product model. The model then is expanded with global structure and master data. The formal description is necessary because all conventional PDM systems and standardized exchange formats such as Standard for the Exchange of Product Data (STEP [ISO 10303]) are based on this concept.

The core of the product model is the product structure, which is shown in the form of a tree structure with extended characteristics. A product is broken down into components and individual parts. The structure ends in parts that cannot be broken down further from a technical point of view. Various characteristics of a product that differ only in detail do not give rise to a complete copy of the structure. Through dynamic integration of the relevant elements in the collectively superposed structural elements, the various characteristics are generated from one structure. Versioning and variants are mapped as additional information in the model.

3.5.3 Process Model

The basic concept of the PLM includes not just the global management and provision of all product-relevant data. The correct handling of information is another significant component. Typically product-related processes are, for example, release and modification processes. In order to allow for IT-based process control, all processes must be established formally in a corresponding process model, taking into consideration the information flow and the responsible parties.

If we look at the basic structure of a process-oriented quality management (QM) system compliant with DIN EN ISO 9001:2000 [ISO 9001], the parallelism of the basic concepts of both approaches is significant. With PLM, as mentioned, the entire process from the product idea to the announcement of a product is modeled, whereas a DIN EN ISO 9001:2000–compliant QM system refers only to the most important production processes.

However, not only execution of PLM in the company can benefit from QM, but the opposite is also true. Certification in accordance with DIN EN ISO 9001:2000 has a process-oriented structure and requires formally described processes. PLM can contribute significantly to the fulfillment of these demands.

3.5.4 Implementation Strategies

With regard to the implementation of PLM systems in companies, reports can be found in the specialized press similar to those seen in the 1990s when ERP systems were implemented. Thus it is not surprising that articles in the specialized press are mainly about abortive and overpriced PLM implementations. Positive headlines about successful PLM system implementations are seldom seen.

Both software manufacturers and universities are striving to overcome this with suitable implementation models [ARNOLD ET AL. 2005].

It can be seen that the implementation of a PLM strategy in a company is principally an internal project. It fulfills all basic conditions that justify project status. The following factors play a part:

- Many, if not all, departments are affected, and their multi-layered demands must be taken into account.
- The internal project team is composed of workers from various divisions.
- External suppliers and service providers are to be integrated.
- Considerable internal time and capital resources are required. These must be scheduled in advance, and employees also must be informed of the plan in advance.

In order to work successfully given these conditions and to carry out the appointed tasks successfully, at least a minimum of project management methods must be observed. These include

- There is an officially appointed project manager.
- The members of the project team are officially released for this task.
- A project structure plan and a phase plan are drawn up within the team based on the requirements and objectives.

Just implementing a PDM system generally fulfills the basic conditions for a project and therefore should be handled accordingly. A pragmatic approach in which you "simply start" regardless of fulfillment of the basic conditions leads almost inevitably to delayed completion and considerable additional costs, if not to the failure of implementation.

Generally helpful for IT implementation projects related to internal structures is a project mentor from general management. The presence of such a mentor at project team meetings and project presentations highlights the importance of the project.

3.5.5 Points of Contact with MES

As will be seen in subsequent chapters, the approaches, basic ideas, and intents of PLM and MES are very similar. The starting point of both systems is a common data model, which on the one hand maps all relevant components and on the other is available to the entire system. The PDM approach thus can be seen as essential for making both a PLM system and an MES possible.

In Fig. 3.6, the entire lifecycle of a product, from idea to scrapping, is illustrated horizontally. As has been shown in detail in Secs. 3.5.1–3.5.4, the PLM approach includes all these phases. The workflow of an MES is displayed vertically. Because of the complexity involved, not all components have been shown.

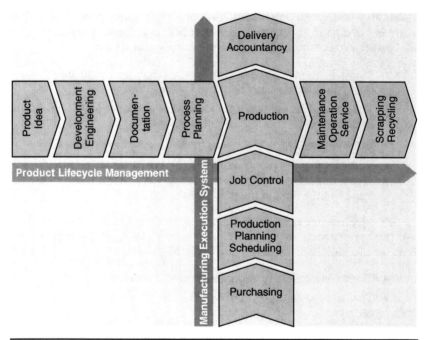

FIGURE 3.6 Concept of product lifecycle management [ARNOLD ET AL. 2005].

As can be seen, there is an overlapping of both systems in the production phase. The following example should clarify this.

On the basis of one or several customer orders that are entered into the ERP system by sales and marketing, a production is planned by the MES with batch sizes, start and end dates, etc. When setting up the machine, the employee discovers that changes must be made to the numerical control (NC) program because of deviations in tolerance for the manufacture of the product. However, these are normally generated on the basis of a computer-aided manufacturing (CAM) model, which, in turn, is derived from a CAD drawing. In order to ensure that the changes will be made permanently, not only must the NC program be adjusted in the order, but the information also must be fed to the design department. This workflow, in turn, is mapped by the PLM.

As shown in the example, not only are there points of contact between PLM and MES, but there are also overlaps between both systems. To what extent these can be resolved through a common system implementation is looked at in more detail in Chap. 11.

3.6 Summary

An analysis of the relevant norms and guidelines shows that there are a number of approaches to an MES. However, closer examination also has shown that all work is based largely on ISA S88 and S95.

The recommendations of the interest groups described provide little new content. The core is formed by the MESA recommendations and their 11 components, which have been specified more finely by the ISA.

The examination of neighboring areas has shown that there is some degree of overlap in the systems. Approaches to simulation, for example, also play an important role in manufacturing and process technology and therefore also should be included as part of the MES.

To conclude this chapter, we examined the PLM concept more closely. The approaches, such as a common database, and to some extent also the functions are similar to those of an MES. There is some overlap between both systems in the area of production.

CHAPTER 4

Core Function— Production Flow- Oriented Design

4.1 Cross-System Cohesiveness

4.1.1 Classification in the Overall System

In Chap. 2 the following division of the manufacturing execution system (MES) functions into three core processes was suggested:

- *Production flow-oriented design.* Data technical mapping of the products (*product definition*) with the production flows (*work plans*) and all resources *needed for production.*
- *Production flow-oriented planning.* Planning of the production process in the form of production orders (hereafter referred to as *orders*) and planning of required resources.
- *Order processing.* Execution (control) of planned orders and acquisition/storage of resulting data.

The three core processes are described in more detail in this chapter (design) and the following chapters [Chap. 5 (planning) and Chap. 6 (implementation)]. The production data model contains a complete description of the product, the actual *product definition*, and, based on this, the resources required and a description of the production environment (mainly machinery and equipment). The degree of detail of the data model arises from the demands made of the planning or execution process. For example, an "optimized setup time" can be executed in the planning process only if the setup times for all machines and equipment per article have been entered in the database; an effective *cost control* can take place only if planning costs for all resources needed and the allocation times in the *work plan* for these resources are present (Fig. 4.1).

53

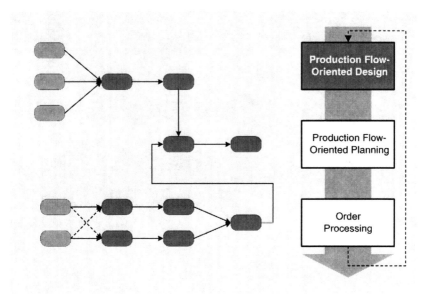

FIGURE 4.1 Classification of the core function "design" in the three main processes of the MES.

4.1.2 General and Complete Data Model

There are few basic normative principles or standardized examples for a universal and uniform production data model. The Instrumentation, Systems, and Automation Society (ISA) defines an *equipment hierarchy* for the hierarchical integration of workplaces/machines into production lines, etc. [ISA S95-1, p. 23]. However, the physical and organizational structure of the production facility is just a small part of the desired production data model. Later in this chapter, a suggestion for a complete data model will be shown that takes all production requirements into account.

The following aspects must be mapped in the MES database using adequate objects:

- *Product definition data with the work plan as the central core.* The *work plan* determines "what" is to be produced and "how" it will *be produced*. In addition, the resources needed—or "with what" the product will be produced—are also classified.

- *Description of the production environment and resources* ("with what" will be produced):
  - Machines, equipment, and workplaces
  - Personnel resources
  - Operating resources such as tools and means of transport

Core Function—Production Flow-Oriented Design

- Materials and preliminary products
- Documents and data on product description and production control
- *System and auxiliary data such as units of quantity.*
- *Data on the execution process* (what is the "result" of the production):
  - Order data
  - Production data
  - Equipment and machine data
  - Performance data (Fig. 4.2)

Even a sophisticated and elaborate data model cannot take into account all possible aspects and features of the various production environments. It therefore must be possible to supplement the basic structure with the basic objects by adding *additional features*. These additional features should be added to the project through parameterization and be saved in the database, just as other project-specific system parameters. This concept gives the MES the required flexibility while still retaining the character of a "standard product" that can be

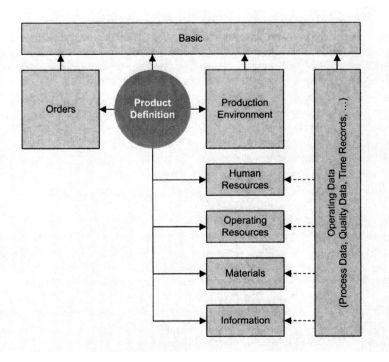

FIGURE 4.2 Macro structure of a general data model for the MES.

updated. The concept of additional features described here affects not only the database, but the additional features also must appear in the system's user interface—without requiring elaborate programming and adaptation.

Another important aspect of the data model is the inclusion of *multilingual applications*. On the one hand, the basic system already should support various languages; on the other hand, all objects and elements that are entered as system parameters must be included in the language concept. The challenge of *multilingualism* affects not only the actual user interface of the system but also other user interfaces such as Short Message Service (SMS) or e-mail as well. Regardless of any permanent "system language" that may be set, it must be possible to send messages to specific recipients in different languages; that is, the language is determined by the recipient of the message.

Especially in the chemical and food-processing industries but also increasingly in discrete piece goods production, it is necessary to *record all user interventions*. For example, changes made to machine parameters, recipes, or article master data should be logged, including the date, time, and the person making the change. In the data model, this requirement is best fulfilled through the *historization of the most important objects*. This means that in addition to the currently valid version of the object (e.g., article master data), all previously existing versions of the object (e.g., the previous definitions of the article master data) are saved. This historization fulfills the requirement for providing supporting documentation and also has the additional advantage that old versions of the object (e.g., a recipe that has already proved itself once in certain environmental conditions) can be "activated" again if needed.

For many objects of the data models, a *grouping of the elements in a hierarchical structure* is needed. For example, this affects personnel, materials, and articles, all of which are allocated to groups. Hierarchical grouping improves the overview and makes it easier to evaluate the data using group-related filters.

4.1.3 Origins of Master Data

Importing Master Data
The basic parameterization of the system always includes some master data, such as the structure and designations of the production facilities or article master data, that may be imported from other systems or a master data management (MDM) system. This requires a flexible import interface. Import should be triggered by the user manually or should be set to ensue cyclically. The imported basic data usually are supplemented by internal MES definitions. The import function must ensure that data sets that have already been imported and supplemented in the MES are not overwritten by the next import. Another requirement for the import concept is that not only can data

from defined sources be mapped 1:1 on target fields but that new data can be created by combining and evaluating the imported data. A simple example here is importing text lists for an alarm system. For this import function, it is useful to be able to derive a "disruption class" (e.g., "emergency off" class) from certain text components (e.g., text containing "*emergency").

Dynamic Creation of Master Data
This concept is closely associated with the import function described earlier. The master data sets are created "on demand," that is, when they are first needed. Importing does not take place as a defined task that creates a certain number of data sets but individually, data set for data set. For the aforementioned "alarm system" example, the text and disruption class of an "alarm message" (i.e., the configuration of this message) must be created when the message occurs for the first time. In this way, the system parameterizes itself independently, and the maintenance outlay is minimized. A condition for this is that the data source, that is, the participant who manages the object, transfers the required information to the MES. In the preceding example, the system must be provided with a meaningful alarm text.

Manual Maintenance of the Master Data
Independent of the previously described automated import modalities, the data must be updated manually. It must be possible to create, edit, and delete all master data sets.

4.2 Data Model for Product Definition

4.2.1 Relevant Concepts
The goal of production is the prompt and cost-efficient manufacture of the *article* with the highest quality—*thus product definition is a core topic of the MES*. The article master data are entered and managed in an enterprise resource planning (ERP) or production lifecycle management (PLM) system, provided that such a system is present. After importing article master data into the MES, the product definition data are supplemented as needed for the MES. In particular, a detailed *work plan* is needed for every article. The work plan determines "what" will be produced and "how" it will be produced. In addition, the required resources—that is, "with what" production will be carried out—are listed. The most important concepts related to this are defined briefly below:

- *Article/article groups*. The *article* is the product that is created in the production work plan using various resources. Various articles in different variants can be manufactured in one production plant. The articles can be sorted into hierarchical *article groups*.

- *Variant. Variants* (or *types*) of articles are manufactured with the same work plan and the same resources as the "master article"; that is, the same operations and work plan can be used for all variants. Differences arise only in details of the operations. Here, different materials, colors, amounts, or work instructions can be entered for variants.
- *Operation.* An *operation* is a defined part of the production process that, as a part of the article-based work plan, is linked with a machine/equipment (operation-machine combination).
- *Work plan.* A series of operations and their links with machines is established in the *work plan.* This sequence also can contain alternative branches and conditional loops. The actual route of the *production unit* (the alternative branch selected) is not determined until the fine-planning stage of the order. The work plan is defined per article and is also valid for its variants.
- *Parts list.* A *parts list* contains the definable individual components of an article in the form of a hierarchical list. These individual components may be raw materials or preliminary products. Internally manufactured preliminary products are also mapped as articles in the MES. Externally manufactured preliminary products ("purchased parts") are treated as raw materials (Fig. 4.3).

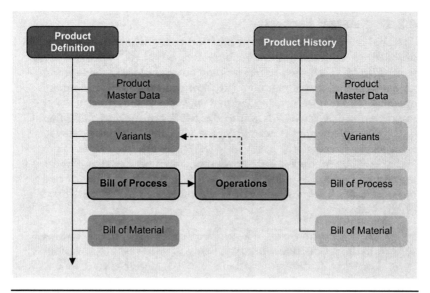

FIGURE 4.3 Main objects for product definition.

4.2.2 The Operation

Overview

The operation is an activity or a process and is a defined part of product development. Some examples are heating, mixing, extracting, rolling, drilling, milling, polishing, transportation, and testing. An operation is independent of a place, such as a machine, equipment, or manual workstation. The link between operations and machines, equipment or workstations (see Sec. 4.3.1) does not arise until the work plan. This link is referred to as an *operation-machine combination* or a *work sequence*. In everyday language, a work sequence or an operation-machine combination is often referred to as an *operation* (Fig. 4.4).

In an operation, a production step is implemented as displayed in the figure with the input of *material, resources,* and *information. Time data* and the required *personnel resources* also depend on the machines, equipment, or workstations detailed in the work plan. Therefore, the data defined in the work sequence (combination of machine/workstation and operation) still can be changed when the work plan is being edited.

Basic Data of the Operation

In the operation, one step of the production process is completed; that is, a defined output is created from a defined input by an action or a process. This process also can have an influence on the production unit. The production unit before the operation (e.g., x meters of real material) can

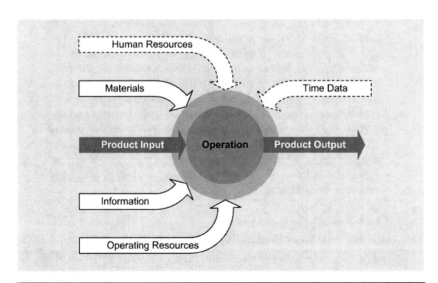

FIGURE 4.4 The operation as a defined part of the production process and the data/resources assigned to it.

be changed (e.g., *y* units of article) by the operation (e.g., "cutting"). It must be possible to map this *transformation of the production unit* (and with it the quantity unit). An operation is determined by the following basic parameters:

- *Name, description.*
- *Activity/process with regard to a general list of occurring activities/ processes.* The selection of an activity/process can affect the remaining parameters of the operation.
- *Quantity input, quantity unit input.* Special forms are found here, for example, with mixing processes in which a so-called initial quantity is determined by the possible volume of the mixer or by the processing of real material.
- *Transformation through an operation.* Taking into account "tolerances" (e.g., changes in volume through the removal of moisture) and "shrinkage" (e.g., cutting).
- *Quantity output, quantity unit output.*
- *Identification of "mixture."* This means that materials used are no longer distinguishable as inputs from the previous operation (e.g., mixing process). After this, statements can be made only about the percentile composition.
- *Parameters for cost monitoring.* If the article goes into a production warehouse after the operation (the warehouse also can be mapped as an operation), specific *storage costs* arise. Storage costs are indicated based on the quantity and time (e.g., costs per piece and hour). In the sense of the idea of activity-based costing (see Chap. 2), the *overhead* of the production is allotted to activities, that is, to operations. The share of the overhead of the product can be defined as the ratio of the number of activities in the operation to the total number of activities. For example, the *material overhead* can be derived from the number of materials used. If, for example, five materials of equal value are contained in the product and one of these materials is used in the operation, the material overhead share is 20 percent. According to the same principle, *quality overhead* (using the number of inspection features), *service overhead* (using the number of operating resources), and other overhead costs can be determined.

Time Data for the Operation

The Association for Work Design/Work Structure, Industrial Organization and Corporate Development (REFA) defines different "time types" and the related evaluation methods within the scope of the company organization. This approach was adopted by many machine data acquisition (MDA) and ERP systems and therefore became a de facto standard. For the requirements of an MES, the following time types are particularly significant:

- t_e = *production time* (also known as *time per unit* or *target time*). We differentiate between quantity-independent and quantity-dependent production times (it is possible to have both types in one operation). If an operation is defined for a transport activity, the production time is equivalent to the *transport time*.
- t_r = *setup time*. In process engineering, the concept of setup time is not used. Here, we speak of *cleaning time* or *preparation time*. The setup time is independent of the quantity produced.
- t_a = order time = $t_r + t_e \times$ quantity.

In some cases, the specifications on "machine cycle time" and "personnel time" are useful as an addition to production time (t_e).

For products with chemical reactions or drying processes, operations may have a *minimum time*. The next operation can be started only after this minimum period has elapsed. One example here is the "maturing period" for dairy products.

Use of Personnel

Personnel requirements are set only roughly for the operation; that is, the *number* of employees required based on time data is provided for every time type (setup and production). In addition, the required qualification (skill) of the personnel and a wage group are also specified. Which and how many staff members are actually employed can be identified and determined in the work plan (in the course of linking the operation with a machine) and in the context of operative order planning, taking actually available personnel resources into consideration.

Use of Material

The use of material is defined with regard to the basic quantity unit of the article (e.g., 1 piece, 1 kg, 100 L, etc.) in the operation. In the execution process, the quantity that is actually used depends on the size of the production unit (e.g., 1 piece, 150 kg, 2,500 L, etc.). We distinguish between the following material types:

- *Raw material/purchased products (externally produced articles).* In every operation, a list of materials used can be defined, with the quantity indicated based on the basic quantity unit. The material is selected via the material number, and a name is selected (with reference to the materials already defined by material management). In process engineering, this list is usually part of a recipe. This means that instead of a list of individual materials, a *recipe* (materials including process description in the form of a workflow) is defined for the *operation*. The operative order planning must check the existing material stock and take it into account for planning.

- *Preliminary products (internally produced articles).* If articles produced by the company itself are needed as "materials" in the operation, a list of articles used also can be defined with reference to the quantity based on the basic quantity unit. The article is selected via the article number, and a name is selected (with reference to the article master data). The planning process (operative order planning) must analyze this "process chain," recognize articles produced by the company itself, and open and plan separate production orders for these articles.

The use of particular materials and articles can be subject to a time limit. This time limit can be indicated using two date specifications "May be used from . . ."/"May be used until. . . ."

Use of Operating Resources

As with material, the use of *operating resources* is related to the basic quantity unit of the operation. It is also possible to list numerous resources per type with reference to the resource management master data:

- *Tools.* For integrating or updating tools, it is often necessary to link maintenance or operating instructions.
- *Means of transport.* Through operative order planning, the capacity of the means of transportation is related and linked to the quantity of the production unit (defined in the order) using the basic quantity unit. The planning process can use this to determine the need for means of transport.
- *Packing.* Procedure for operative order planning as with means of transport.
- *Measuring equipment.* Operational instructions are often needed for the use of measuring equipment.

Classification of Information and Documents

Work instructions, process instructions, and test plans (see Sec. 4.3.5) can be linked with the operation. For each document, it is necessary to distinguish how the information should be used:

- *Informative character.* Information and directions for the workers can be viewed optionally in the execution process. The worker receives notification of the information but is not required to read it if he or she is familiar with the content.

- *Information that must be acknowledged.* This information is always displayed in the execution process. The worker must acknowledge the information.
- *Execution of an activity and acknowledgment.* The information is always displayed in the execution process. The worker must carry out the instructions and then acknowledge. Test plans or rework protocols always belong to this group.

Creation of Variants in the Operation
The possible variants of the operation considered here are indicated with reference to the article variant list (part of the article master data). These are generally variants in materials, such as color variants of an article or raw materials from different suppliers. The variants therefore also can have an effect on operating resources and documentation.

4.2.3 The Work Plan

Overview
The work plan (bill of process) is the central control instrument of production. In it, the individual steps of production are defined for each article in the form of *operation-machine combinations,* hereafter referred to as *work sequences.* The following data define a work sequence:

- *Sequence number.* Work sequences are defined clearly within a work plan by means of a sequence number. Intervals of hundreds (sequence number 100, 200, 300, etc.) are used in many systems and allow alternative or parallel work sequences to be integrated easily into the numerical scheme.
- *Operation.* The operation defines the "action" and/or the "process" (what is conducted in the production step) and all allotted operating resources (how the production step is carried out). The operations have either been defined prior to the creation of the work plan or can be created in the course of compiling the work plan.
- *Machine/system/workstation.* The link with a machine, a system, or a manual workstation (hereafter referred to collectively as *machine*) defines the *location of the operation,* that is, where the production step is carried out. Operations can be carried out *alternatively* on several machines (e.g., either machine A or machine B as an emergency strategy) or also in *parallel* on several machines (e.g., both machine A and machine B in order to increase capacity). This alternative or parallel processing is mapped in the work plan as a specially marked work sequence. The respective alternative or parallel machine

Work Sequence Number	Operation	Machine	P	A	Of Work Sequence Number
100	Cutting part A	Band saw			
200	Milling part A	Processing center 1	P		100
210	Milling part A	Processing center 2	P		100
300	Drilling part A	Processing center 3			200
310	Drilling part A	Processing center 4			210
400	Grinding/polishing part A	Processing center 5			300
500	Preassembly 1 part B	Assembly area 1			
600	Preassembly 2 part B	Assembly area 2			500
700	Assembly part A + part B	Assembly area 5			600
800	Testing	Quality control site 1		A	700
810	Testing	Quality control site 2		A	700

TABLE 4.1 Example of a Work Plan with the Basic Elements in Table Form

can be selected in the operative planning system or not until the execution process, depending on the circumstances.

- *Specifying the relationship to predecessor/successor.* In order to map production, details are still needed about the order of the work sequences. One simple method for mapping the order is the specification of the respective predecessor (Table 4.1).

Specifying Details on the Work Sequence

In addition to the basic features described earlier, some details are still needed to complete the work plan. These features depend greatly on the requirements of the respective production environment and therefore must be suitable for configuration. Here are some often needed details:

- *Use of staff.* Personnel requirements per time type, that is, for setup and manufacturing, can already be roughly determined in the operation. These details must be specified more precisely in the work sequence by linking them with a real machine.
- *Ratio of quantity to next sequence.* Normally, input quantity is the same as output quantity; the quantity ratio is 1:1. However, changes may arise in the quantity ratio through mechanical or

chemical processes or through shrinkage. The output quantity then is given as 0.95 (5 percent lower), for example.
- *Start quantity.* The start quantity indicates when an operation can begin with regard to its predecessor. This information is important for operative planning and for determining the overall processing time. For example, the next work sequence can start in a piece goods plant when a transport container has been delivered with parts from the previous process. Thus, in this case, the start quantity would be the same as the capacity of the transport container in pieces. Cycle times can be reduced considerably by "overlapping" operations.

Planning Costs

After the work plan has been drawn up, the planning costs for the article are set. The overall planning costs can be calculated by adding all *resource costs* (i.e., staff, machine, material, and resource costs) and the *overhead* (in accordance with activity-based costing regulations) of work sequences. These planning costs can be saved with the article master data and also issued as a detailed report.

Tools for Creating the Work Plan

Defining operations and allocating all necessary resources and information, as well as creating the actual work plan, form a complex parametric exercise. In order to make this exercise as user-friendly as possible, a sophisticated user guide (for defining the operation) or preferably, graphic editors (for creating the work plan) should be used, or it should at least be possible to display the work plan created as a graph (Fig. 4.5).

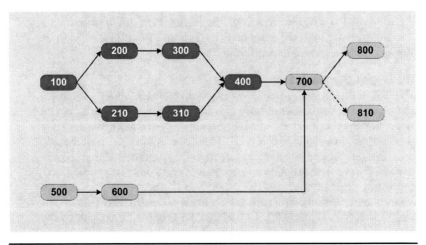

FIGURE 4.5 Graphic representation of a work plan (with reference to the example shown in Table 4.1).

4.2.4 The Parts List

The *parts list* is a hierarchical classification of the product into individual components (raw materials and preliminary products) that can be derived from the work plan. The entire production process and the required material resources are defined, as described earlier, in the individual operations. The parts list arises from the sum of the materials and preliminary products used. The hierarchy of these materials results from the production flow depicted in the work plan. This means that the parts list must not necessarily be maintained as an independent object in the data model but can be created online as a report, for example.

4.2.5 Change Management and Product History

The article is subject to changes throughout its lifecycle. These changes also must be indicated in the *product definition,* possibly with the aid of a PLM system where this is present.

Normally, the basic structure of a product is largely retained with the work plan over the lifecycle (greater changes lead de facto to a new article). Changes are much more likely to arise in the details of the operations, for example, in the materials used or the process instructions applied. Change management above all must record changes in operation definitions and archive all previous versions. Older versions must remain accessible. Not only is this necessary for traceability and product liability, but it also makes it possible to revert to "older" versions of the article and produce these.

4.3 Data Model for Resource Management

4.3.1 Description of Production Environment

Figure 4.6 shows a description of the production environment as a logical structure with the main objects, machines/equipment, interim storage/buffer, terminals, and time models.

The Logical Structure of Production

The ISA defines an *equipment hierarchy* with elements as a hierarchical structure. It establishes not only hierarchy of production but also an allocation to the level model (e.g., company management level or production management level). This perception is problematic because, on the one hand, different terms and hierarchical levels are used in different production environments (e.g., discrete production/continual process) and, on the other hand, one perception (logical and/or geographic) of the company is often not sufficient. This means that although it is advisable to use the ISA scheme as a support, the number and names of the hierarchical levels should be set flexibly. Along with a logical/geographic perception, other viewpoints also should be included. It therefore should be possible to classify machines

FIGURE 4.6
Description of the production environment.

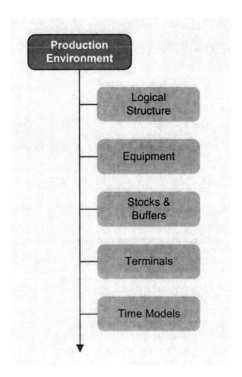

and equipment not only according to the factories and production divisions but also, for example, according to maintenance divisions. The maintenance employees thus have their own perception of the production, one best suited to their work (Table 4.2).

The machines and equipment, production stocks and buffer, and time models described below are independent objects of the data model, whose elements can be linked with the logical structure.

Machines, Equipment, and Workstations

The machines, equipment, and workstations of production are primary objects. The defining characteristics result from different perceptions:

- *Basic data*, such as name, description, manufacturer, serial number, inventory number, cost center, and image
- Data to *describe production characteristics*, such as maximum capacity, machine hourly rate, lead time until operational readiness, average setup time, etc.
- Data for *servicing*, such as guarantee period, maintenance cycle, reading for condition-based maintenance, or energy consumption/connection load

Hierarchy Level	ISA Term	Relationship to Level Model
0	Enterprise	Level 4—Enterprise management
1	Site	Level 4—Enterprise management
2	Area	Level 4—Enterprise management
3	Process cell	Level 3—Production management
3	Production unit*	Level 3—Production management
3	Production line	Level 3—Production management
4	Unit	Level 3—Production management
4	Work cell	Level 3—Production management
5	Lower-level equipment used in batch operations	
5	Lower-level equipment used in continuous operations	
5	Lower-level equipment used in discrete operations	

*The *production unit* as a part of the production structure should not be confused with the concept *production* unit that was defined previously in this chapter.

TABLE 4.2 The Equipment Hierarchy According to the ISA with the Recommended Systems for Level Management [ISA S95-1, p. 23]

Subordinate to the machines and equipment is the setup matrix. This structure shows how long the sum of the times for a shutdown procedure (of a defined, produced article) and setup procedure (as preparation for the production of the next article) is. This time generally is referred to as *setup time* and is an important parameter in many production environments for fine planning and sequence optimization (Table 4.3).

If the setup time is not subject to any particular fluctuations owing to different articles, an average value can be entered directly in the master data for the machine/equipment.

Transport Routes, Production Stocks, and Buffers

The transport routes, interim stocks, and buffers integrated into the production flow (referred to as *production stocks* in the text below) must be known in the MES for functional fine planning to work. Only by including the production stocks can true time and resource planning

Setup from/to	Default	Article A	Article B	Article C	Article D	Article E	Article F	Article G	Article H
Default	—	600	660	660	X	X	X	X	X
Article A	600	—	600	X	X	X	X	X	X
Article B	660	660	—	X	X	X	X	X	X
Article C	540	540	X	—	X	X	X	X	X
Article D	X	X	X	X	—	X	X	X	X
Article E	X	X	X	X	X	—	X	X	X
Article F	X	X	X	X	X	X	—	X	X
Article G	X	X	X	X	X	X	X	—	X
Article H	X	X	X	X	X	X	X	X	—

TABLE 4.3 Schema of a Setup Matrix with Sample Values in Seconds

be carried out; they can be used to decouple various production areas or as interim stocks for partial products. In both cases, the sequence of the articles (i.e., which article in which quantity), the maximum and minimum stock (in order to avoid breaking off production in normal conditions), and the transport times must be known.

Terminals

Terminal does not describe a "standard client" of the MES (such as a Web client) but rather an on-site terminal (at the machine/equipment, usually in the form of a rich client). These terminals usually have functions that are determined to a great extent by the on-site conditions. This means that terminal A and terminal B will have different user functions. The terminal's required functions are saved in the database with the description of the associated machine/equipment and a clear reference. After hardware replacement, the original function can be restored immediately. The computer name can serve as a reference, for example (IP addresses are less suitable as references).

Time Models

Time models make it possible to classify the total time into *production time* and production-free time, for example, with the aid of shift models. A production day can be different from a calendar day—for example, it can last from 06:00 a.m. until 01:30 a.m. on the following day—and is defined using any number and order of shifts, in other words, by a shift model. The shifts, in turn, can contain arbitrary "breaks." For planning production, a series of production days is determined in a weekly or monthly scheme in advance. The system must be able to recognize fixed and movable public holidays for the specific country.

In the flexible working world, time models also must be implemented very flexibly. In an extreme case, different time models for each machine can exist even at the production unit level (see Table 4.1). The link between the time model and logical unit from the production structure determines the *production time* for this unit.

Determining production time has far-reaching effects for order planning (which resources are available when), for time recording in general, and also for capacity assessment for machines and equipment (e.g., technical availability and overall equipment efficiency). Therefore, precise maintenance of the time model is of great importance for the overall data quality of the MES. Changes in the time model are always possible, and it must be possible to input these at short notice—in extreme cases, during the current shift. For example, if an *all-hands meeting* is defined as an additional break in the system, the target values (lower quantities) as well as the basis for calculating key performance indicators (KPIs), which depend on the *net production time,* change as a result.

It should be possible to map company-specific time definitions, for example, a "business year," that differs from the calendar year and therefore "delivery weeks" instead of normal calendar weeks.

4.3.2 Production Personnel

Figure 4.7 shows the basic elements of human resources management.

The master data set of an employee contains at least the following data:

- Name, surname, birth details, and entry date into company
- Personnel number and ID for automatic identification

This master data set usually can be imported from the ERP system or a special module for human resources management. In the MES, additional data are needed for calculating performance payment details and personnel planning. These details are managed in separate tables and are linked to the employee master data sets.

The modalities and rates for performance payments (e.g., piecework, group piecework) are indicated via the wage groups. A list of skills in conjunction with personnel planning management makes it possible to carry out fine planning for orders with regard to the available personnel resources. There is an $n{:}n$ relationship between employee and skills; that is, each employee can have several skills. Personnel availability can be mapped in a special planning instrument

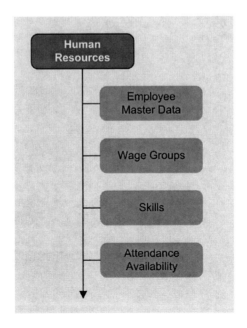

Figure 4.7 Basic elements of human resources management.

(personnel management). Attendance is planned taking holidays and illness into consideration.

4.3.3 Operating Resources

All operating resources to be managed through the MES as a condition for planning and fulfilling orders must be mapped. Operating resources are usually defined by type (e.g., means of transport, lattice box) and an available number (e.g., 100 lattice boxes in circulation). For important resources that should be identified clearly for production data recording, the ID for each individual part also must be administered. An operating resource is linked with an article in the work plan. This establishes which resource in which quantity is needed for production in an operation.

The following operating resources are needed often and therefore should be mapped in the standard scope of the product. Any other resources needed also can be defined (Fig. 4.8).

- *Tools.* For complex tools such as screw spindles, it is necessary to ensure that a particular tool can be used in various stations (e.g., the screw spindle is installed in a different station after successful maintenance). This means that the allocation between station and tool must be saved, and the operating hour meter must be linked to the tool.

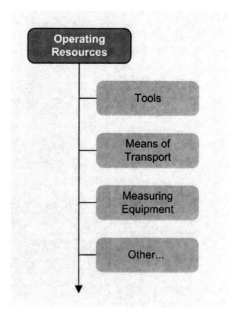

FIGURE 4.8
Frequently used operating resources.

- *Means of transport.* For means of transport, the capacity (what quantity of an article can be loaded) must be entered as master data for each means of transport, and the current occupancy must be entered. The operating resources then can be considered in order planning.
- *Measuring equipment.* The current use and availability of the measuring equipment must be managed per "measuring equipment" type in the MES.

4.3.4 Materials and Preliminary Products

Both externally produced preliminary products (known as *purchased parts*) and raw materials are summarized below under the concept *material*. The materials needed at production should be managed per production site. Each material must be labeled clearly with a *material number* or a *material code*. In addition, the following master data must be administered for the material:

- Name, abbreviated reference, description
- Allocation to a material group
- Supplier, quantity unit, calculated costs, and stocking costs per quantity unit (Fig. 4.9)

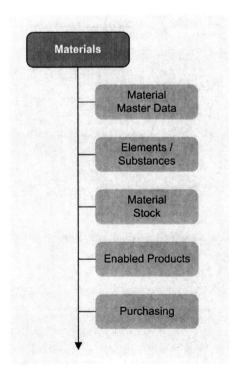

Figure 4.9 Data model for resource management in the MES.

For materials used in the foodstuffs or pharmaceutical industry, it is absolutely necessary to document in the MES all ingredients (the substances contained in the material), with the proportions provided by the supplier. This *material summary* is important for creating recipes, for achieving the required product quality, and for product documentation (recording product data). An $n{:}n$ relationship should be set up between the material and its components indicating percentages. This means that every material can be composed of n ingredients.

Since material is the only resource that is subject to continual use, updated information on supply is necessary to provide an overview of available, reserved, and ordered material at all times. This means that the stock on hand for a material must be managed through the MES. In addition to the storage location, existing packing units, dimensions and weight, details on shelf life, and a minimum stock must be managed. "Booking out" parts from the warehouse is carried out via *material usage* in a production operation. "Booking in" of new material is done using the incoming goods function. The total warehousing costs can be calculated and optimized by the MES using the current stock and the warehousing costs of the material.

Packaging is a special type of material. For packaging, the capacity (what quantity of an article is packed) and the current availability (how many packages are still in stock) must be saved as master data per *packing type*. This means that this resource can be taken into account in order planning.

Not every material in a material group is necessarily suitable for producing every article. Therefore, a *material release list* may be necessary for recording which materials may be used for which articles.

Material acquisition (see also Chap. 5) must be implemented in the MES at least in the form of "order suggestions" that are transmitted to an ERP or merchandise management system. Only thus can stock be optimized (goal: reducing warehousing costs; see Chap. 8) and timely order processing guaranteed. If overall order processing is carried out by the MES, supplier data (e.g., best ability to deliver, lowest price, etc.) and a price history with the current price will be needed.

4.3.5 Information and Documents

Overview

Figure 4.10 shows a data model on information management in the MES.

The MES must pass on all information needed within the production flow to people (mainly workers) and machines/equipment on time. This information consists mainly of *process instructions* and *work instructions*. In addition, documents that control or accompany the

Core Function—Production Flow-Oriented Design

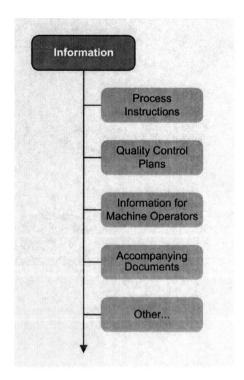

FIGURE 4.10 Data model on information management in the MES.

production flow (e.g., accompanying documents) must be created and managed.

Since this information is generally relevant for quality, the respective documents or data sets also must be stored in a change protocol along with versions and changes with date/time and respective user.

Work Instructions/Worker Information

Work instructions are aimed at workers and provide them with instructions and regulations for carrying out a particular work step. The instructions are related to the article being produced, on the one hand, and on the other, they usually refer to a defined operation. The MES therefore must provide the information for the respective article on the correct operation (machine or equipment). The information is displayed either on an MES terminal directly on the machine via a permanently installed large-screen display or via portable devices. The instructions may be displayed as graphics or text. For more complex work instructions, interactive documents are also useful; for example, the worker can request the instructions for the next step after completing a work step.

Work instructions can describe the following activities/processes:

- Assembly
- Manual settings of machine parameters
- Manual settings for recipes
- Monitoring functions, for example, for machine parameters through comparison of target and actual values or targets for a statistical process control (SPC) system

Process Instructions

Process instructions are passed on directly to automatic control systems in production. Like work instructions, they are related to the article currently being produced and generally affect a defined operation. The data must be present in a format comprehensible for *production control* (e.g., as a CSV, TXT, or XML file). In addition, the data can be displayed to workers for control purposes (see "Work Instructions").

Process instructions may describe the following functions/processes:

- Automatic settings of machine parameters.
- Automatic settings for recipes.
- Monitoring functions, for example, for machine parameters through comparison of target and actual values or targets for an automatically run SPC.
- Control programs such as NC programs, robot programs, or recipe/batch programs, hereafter referred to collectively as *programs*. The programs are created in external systems (e.g., as the result of a simulation process with CAD tools) or through direct on-site programming but then should be managed through the MES and passed on to *production controllers* as process instructions. These programs therefore must first be transferred to the MES. This is done either through a *teach-in process* (the MES takes on a "tried and tested" program from production control via the existing interface) or by transfer via data-storage device.

Test Plans

Test plans are a special form of work and/or process instructions. A test plan contains a variable number of features that belong to a *feature class* and to which a unit of measurement is allocated. We differentiate between *measurable features* and *attributive features*. The measurements of measurable features are recorded and monitored within a statistical quality control (SQC) system. This also applies to attributive features, which are evaluated in the form of a mark, a text, or an error type that has occurred.

The following data are important for automatically testing the feature with regard to its "capability" in the process:

- Tolerance lists (These limits are established by the production manager/quality manager and must be strictly adhered to.)
- Details on the expected distribution form
- Size of sample
- Intervention limits (When intervention limits are exceeded, measures must be taken to stabilize the process.)
- Details on testing frequency (Details such as "every hour," "every tenth roll," or "four times per shift.")
- Indication of the measuring/testing methods used (see Sec. 4.2.4)
- Details on measures if limits are exceeded (The measures can be saved in separate data sets/files in which the individual measures are recorded along with execution steps.)
- Optional details on warning limits and standard deviation
- Optional details on documentation type

Accompanying Documentation

Documents that accompany the production flow are, for example, *accompanying documents for goods* (accompanying a production unit) or internal *delivery notes*. These documents usually have only limited validity within a production location. The MES must create and print these documents at any time when an order is released. In the case of loss, it must be possible to create the relevant document again on request at any time. In *paperless production* (which should be encouraged), these documents are replaced with internal data flow in the MES and tailored to on-site visualization. However, paperless production control requires that the articles can be identified at any time in the production flow. Technical aids here are, for example, barcodes or RFIDs (see Sec. 6.3.2).

4.4 System and Auxiliary Data

Figure 4.11 shows system and basic data in the MES.

Among the *system data*, user administration should be especially emphasized. A detailed allocation of rights is necessary for all users. Authenticating the user and verifying the password also can be carried out in conjunction with an external system (e.g., active directory, LDAP server).

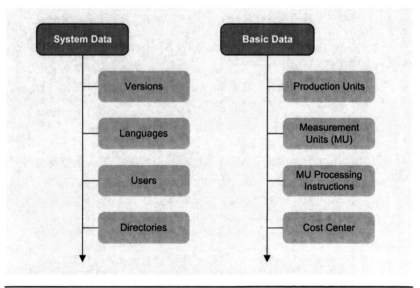

FIGURE 4.11 System and basic data in the MES.

The *basic data* describe the basic features and parameters of the system, for example, which *production units* are present and what units of measurement are needed. We distinguish between simple (e.g., m or kg) and complex (e.g., m/min or kg/m^3) units of measurement. The *basic quantity unit* plays a particular role among the units of measurement.

Production Unit
The production unit is a defined quantity of the article, independent of the actual order. Examples are lot (in serial production) or batch or load (in process engineering).
 The basis for production in lots or batches lies either in process engineering requirements (e.g., if a process has been optimized to the size of a reaction vessel) or in economic aspects (e.g., reduction of setup times). Thus a production unit can include several orders, but an order also can extend across several production units.
 The production unit is identified independently of the order (e.g., by lot number, serial number, charge number, or unambiguous article ID), linking it with the order(s) derived from it. Thus it serves above all as a basis for product tracing and the recording of all product-specific data. If the basic quantity unit is "one piece" (discrete individual production), the production unit is equivalent to a particular article, which as a rule is also clearly identifiable across all work processes.

> **Basic Quantity Unit**
> This quantity unit refers to the article to be produced and is used as a reference value for all specifications and calculations in the operations. For example, the time specifications, the number of preliminary products used, and the required quantity of raw materials are always based on these basic quantity units. In discrete production, the basic quantity unit is one piece. In process engineering, this unit can be "10 kg" or "100 L."
> In complex processes, the basic quantity unit can change in the individual operations. Thus, for example, a mixture of x kg can be transformed into an intermediate product in the form of a real good of x m in length in an extruder. In subsequent processes, a transformation of this intermediate product into individual parts (pieces) is conceivable. Thus it must be possible to define the basic quantity unit or a transformation of this unit for every operation.

4.5 Order Fulfillment Data

4.5.1 Orders

The data model for order fulfillment contains the order data, data from production data acquisition, and performance data derived from this (Fig. 4.12).

Customer orders are usually mapped in an ERP or merchandise planning system. The production orders are derived from these; production orders are referred to in general as *orders* in this book. A customer order can lead to several production orders. Furthermore, a production order can contain several customer orders (or items from these customer orders). In the absence of an ERP or merchandise planning system, the relationship between customer and production orders can be managed by the MES.

An order is essentially defined by an *article* (what is to be produced), the relevant *quantity* (how much is to be produced), and a *delivery date* (by when the good must be produced). In order to facilitate the allocation of orders to equipment/machines (work sequences), the orders also can be mapped with respect to operations.

4.5.2 Production Data, Operating Data, and Machine Data

During the execution process, data arise that must be allocated to the orders and archived. The following data can be distinguished:

- *Order data*. In the course of order fulfillment, in particular the currently produced quantities must be recorded and compared with the target quantities from the orders. Collecting and reporting data is done per work sequence, that is, with reference to

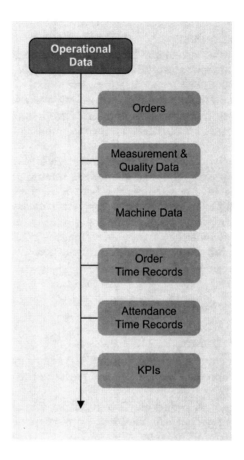

FIGURE 4.12 The data model for order fulfillment.

an operation. The actual duration of the work steps with allocation of the individual personnel and machine resources is also included in the order data. Personnel and machine utilization thus can be calculated per order (based on the article).

- *Material usage.* The identity and quantity of the materials and preliminary products used must be recorded consistently and linked with the article. These data can be traced on the basis of a clear identification (e.g., serial number of the article) or by production date.

- *Measurements and quality data.* Various measurements providing information about the quality of the product also must be recorded and archived. As with material usage, traceability is ensured by the clear identification of the article or a time stamp.

- *Machine data.* Machine settings are treated as *measurements/ quality data* where they are relevant for the production process. In addition, machines also provide data for alarm management and preventive maintenance.

- *Personnel time.* Using personnel *time recording*, the attendance of staff is recorded on the one hand, and on the other, the actual activities (on which machine/equipment the employee has worked for how long on which order) are recorded.

4.5.3 Derived Performance Data and Figures

From the data described in Sec. 4.5.2, the system derives figures for performance assessment, which are called KPIs. We need to differentiate between values calculated online (figures that are calculated continually and reflect the current production status) and archived values. Archived figures are measured at the end of a defined period (e.g., a shift or a day) and then archived. Overall production status is recorded using compressed figures. Individual values that result in these figures can be removed from the system after an adjustable period of time. For example, isolated disruption events in a shift need not be saved for a long period of time because compression in the form of disruption duration and number of disruptions is made for every shift.

4.6 Summary

An MES as a tool for the production management level (level 3 in the ISA model) requires complete *product definition;* an ERP system (company management level—level 4 in the ISA model) requires only parts of it. The logical conclusion is that the entire administration of product data can be assigned to the MES. Through transparent structures and suitable software technology, the MES *product definition data* must be available to all other applications.

At the core of the data model is the *work plan*, which describes the production process of an article as a sequence of activities/processes with all required resources. Since there are large differences, both technological and structural, between various products, a flexible and expandable universal data model must be developed. In particular, aspects of mixed production with process engineering and discrete production steps and the different quantity units that result must be taken into consideration.

CHAPTER 5
Core Function— Production Flow- Oriented Planning

5.1 Integration within the Overall Process

In Chap. 4 we looked in detail at the product data model. Based on this complete, consistent data model, which describes both the article to be produced with its work plan and the available resources (i.e., workforce, machines, and material), *operative and production flow-oriented planning* (production flow-oriented planning) is carried out (see Fig. 5.1).

Today, master data relevant for the planning process usually are maintained only rudimentarily in an upstream enterprise resource planning (ERP) system and therefore must be completed in the manufacturing execution system (MES). Since multiple mapping of data should be avoided, it should be a future task of the MES to administer the entire range of master data. This approach then would comply with Instrumentation, Systems, and Automation Society (ISA) directives (see Sec. 3.2.1).

5.2 Order Data Management

The basic tenet "maintenance and data management in one system only" also must apply to order data. Double data management or manual transfers lead to additional costs and errors. In general, the order data are acquired via the ERP system and then are transferred to the MES ahead of schedule.

The interface between the ERP system and the MES depends primarily on the performance and use of the systems implemented. However, the MES requires at least the following data to carry out production flow-oriented planning, and these master data usually are maintained in the ERP:

84 Chapter Five

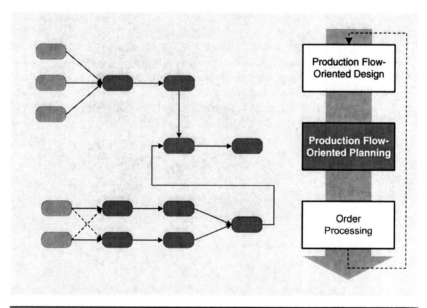

FIGURE 5.1 Core element 2: Production flow-oriented planning.

- Article (with commercial data)
- Machines (with limited information)
- Personnel (possibly with employee qualifications)
- Work plans (usually in a rudimentary form)
- Part lists (often in addition to work plans)

On the basis of the customer orders that are maintained in the ERP system, the MES creates production orders. It is possible to merge several customer orders in one production order. With larger customer orders, however, a customer order may be processed in several production orders. This decision is made in the MES. Accordingly, it is necessary to transfer the customer orders to the MES. At a minimum, the following order data are needed:

- Unique customer order number
- Unique article identification
- Quantity with unit (e.g., pieces, liters, kilograms)
- Date (the earliest possible production date depending on material availability or the planned delivery date)

With regard to the delivery date, it is advantageous to distinguish between the earliest possible date, a request date, and a latest date. Awareness of these details increases the flexibility of fine planning.

Customer data (e.g., customer number, name, and delivery address) also should be imported from the ERP system. They enable the MES to create labels for packing units should they become necessary.

The *core task* of MES is to put the orders transferred from the ERP system into an *optimal sequence*. Here, there are many different approaches, from simple manual planning to a "planning board" to fully automatic planning based on defined criteria such as planning to minimize setup times. If there are several orders for the same article, it must be possible to create collective orders, and the time period–related quantity needed should be calculated by means of equalization processes (e.g., *Heijunka*).

Heijunka
This is a concept from Japanese production: harmonization of the production flow through quantitative production balancing, which avoids queues (i.e., idle periods and transport times). Factory production is replaced by flow production (i.e., continuous-flow manufacturing) with short transport routes and complete processing. This is of great significance, especially in view of complex, multistage production. The bottleneck sector limits the company's success (law of balanced planning) and also creates wastage for all other sections (in the sense of delays, stock, etc.). [SYSKA 2006]

Using simulation, an MES should run various scenarios with regard to variants, quantities, and dates. Figure 5.2 shows an overview of the functional sequence of individual activities [customer relationship management (CRM), product data management (PDM), and advanced planning and scheduling (APS)].

5.3 Supply Management within the MES

5.3.1 Demand Planning

The central task of material requirements planning is to ensure material availability through supply management. This means procuring the necessary required quantities within the company and for sales at the proper time.

This also includes monitoring stocks and in particular creating procurement proposals for purchases. Here, demand planning strives to find the optimal path between

- Best possible supply readiness and
- Minimization of costs and capital lockup

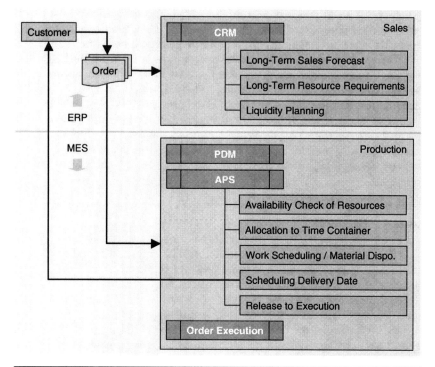

FIGURE 5.2 Functional sequence from customer order to production release.

5.3.2 Material Requirement Calculation

As a first step, the planned resource requirements are calculated with the aid of an operative planning forecast and coordinated with the stock and availability situation. On this basis, decisions are made for a reservation, for ordering, etc.

If data distribution is conducted in accordance with ISA guidelines (see Sec. 3.2.1), work plans and therefore part lists are contained in the MES. In order to be able to access the stock of all stock types directly from supply management, stocks must be listed in the MES.

5.3.3 Material Disposition in the MES or ERP System

In today's systems, material disposition generally takes place at level 4. In future production management systems, production will plan in the short-term on the basis of a real situation. New-order situations can be calculated very quickly using the operative planning tools.

If orders must be made at short notice, the call-off order ideally is sent to the supplier by an e-mail or is made directly available to the supplier via a business-to-business (B2B) platform on the Internet. The supplier transmits order confirmations and delivery dates in the same manner. Since the product data system will be managed at level 3 in future systems, the latest data for work plans, together with their parts lists, are always available to the planning department.

For calculating material requirements, required *preliminary products (articles) from in-house production* also must be calculated and their production planned. Based on the actual production order, "subcontracts" can be generated automatically by the MES for production of the preliminary products and assimilated into the sequence planning.

5.3.4 Incoming Goods

In the incoming goods function, an incoming goods number is provided automatically with reference to an open order. Then the individual order positions are checked. Since deviations may arise during quantitative checking of the delivery positions, the actual incoming quantities, which are compared with the order quantity, should be calculated. If an order has been made based on packing units, every delivery unit should be recorded and managed according to stocks.

There may be some possible overlapping in quality management in incoming goods inspections. In this case, a material booking should be carried out via the MES. For every recorded delivery unit that is brought into a warehouse, accompanying documents should be created to facilitate identification of the goods.

If an incoming goods inspection in the sense of quality control is needed for the material, this must be visualized through the MES. Either a direct, on-the-spot inspection can be carried out or a laboratory order can be generated automatically for the material. The material is released for production only after positive inspection. For the testing process, we refer to the inspection procedure used in production, which generally agrees with the process in the incoming goods section.

After the quantitative inspection and quality tests, material booking must be transferred to accounting. Under certain circumstances, specification of a cost center and a cost unit may be necessary.

5.3.5 Interaction between the ERP System and the MES

As we have shown, there is some overlapping of tasks between the ERP system and the MES in terms of material management. In this regard, there are various approaches for allocating these tasks:

1. The best option is a service-oriented architecture (SOA). Here, one tool, for example, the MES, uses the material booking of another tool, in this case the ERP system, as a service. This approach has defined interfaces and exchange formats. The service is started through the handover of an ID with a corresponding return value.

2. Another suitable solution is saving the data using a data warehouse. This allows for a global view of heterogeneous and distributed data in which the data relevant for the global view from the data sources are merged into a common, consistent data set. Thus the content of a data warehouse arises by copying and formatting data from different sources, in this case an ERP system and an MES.

3. The last option involves managing the tasks in two separate tools. The result is that the data sometimes are saved in duplicate, and therefore, more maintenance is required. Errors in data management cannot be avoided.

5.3.6 Material Warehousing Costs

The quantity on stock (inventory) is to be assessed regarding accumulated storage time using the warehousing cost rate. The resulting warehousing costs are to be displayed. They must be available to the MES for detailed planning and optimization. Inventory includes stocks (raw material), stocks (semifinished goods), and stocks at production, also known as *work in process* (WIP).

5.4 The Planning Process

5.4.1 Planning Objectives

One task of an MES is to bring the orders transferred from the ERP system (generally a pool of orders for a defined time period with a fixed supply) into an *optimal sequence* using suitable algorithms. What the optimal sequence is depends on many frequently changing factors even within one production facility, owing to adjustments in company policy, for example. However, two important aspects are needed to varying extents in sequence planning of every production:

- The customer viewpoint, in which adherence to delivery dates and quality are at the forefront. The priority of the individual customer orders generally is determined by the production planner (order preparation). Rules for this can be entered in the MES.

- The cost-oriented viewpoint (e.g., optimization of setup costs, resources, and warehousing costs or a mixture of different factors) to minimize production costs. The optimal sequence of orders is determined by the MES according to internal rules.

An MES should use simulation to calculate and display different situations with regard to variants, quantities, and dates. Fully automatic sequence planning is possible for production processes with few boundary conditions and clear rules. For more complex planning processes, a decision on which variant is actually produced is made by the process-planning divisions based on suggestions.

5.4.2 The "Updated" Work Plan: Condition for Optimized Planning

Years of experience in the analysis of work plans have shown that a large number of targets (mainly planning times for production steps, setup, cleaning, etc.) differ significantly from the reality. The reason for this is that these targets are set at a particular point in time before the start of production (e.g., based on estimates and time studies) but are no longer checked for accuracy during the production itself. However, actual production times often change considerably through continual improvement processes.

An important requirement of the MES planning function can be derived from this: the MES should have a tool that compares the time targets from the work schedule with the actual times and then adjusts the target times. The work plan thus can be adjusted to reality by an automatic regulatory system, leading to more reliable and precise sequence planning.

Recording the actual times can be performed simply and cost-effectively by means of a production data acquisition (PDA) system, which is often already present or should be a module of the MES. In these systems, order-related and thus article-related actual times are recorded. The statistically calculated average vales of these times then can be compared with the target times, and deviations can be analyzed.

5.4.3 Work Scheduling

In existing planning systems, we often encounter the error that all orders are processed in one order pool, resulting in unacceptable calculation periods when there is a large number of orders. In particular, trying to squeeze in a "priority order" can cause considerable problems because it is then no longer possible to reschedule within an acceptable time in order to have an executable plan for the current shift.

A suitable approach before the actual planning is to divide the quantity of orders into defined time periods (e.g., a day or calendar week), known as *time containers*. This approach leads to a better overview and shortens the calculation times for planning. In the simplest case, dividing the orders into the time containers can be done manually by order preparation.

The problem of sequence planning always arises when several orders compete for scarce production resources. It is therefore relevant to both order-oriented individual production and type and series production. Only with mass production, with its strongly specialized production facilities that are concentrated on a particular type of product, is operative sequence planning seldom necessary. The complexity of assigning tasks can be seen by viewing various boundary conditions. These are multilayered and are often contrary to the planning goal. Here is an overview of frequently occurring influencing factors:

- With regard to the customer order:
  - Delivery date
  - Delivery quality
- With regard to the product:
  - Alternative work plans
  - Alternative part lists
  - Setup costs depending on the sequence
- With regard to the production process:
  - Minimum or maximum intervals between the process steps
  - Transport times
  - Waiting times (e.g., cooling or maturing process)
- With regard to production resources:
  - Current resource allocation
  - Availability of means of transport and other resources
  - Cleaning and maintenance times
  - Availability of quality assurance resources (e.g., test stations, laboratory capacities, etc.)

Therefore, for the selection (and subsequently, the parameterization) of an MES, it is important to clarify these influencing factors. Only when these factors and their priority are clearly known can effective sequence planning be carried out, leading to an improvement of the entire production system.

For sequence planning for a larger order pool with the goal of *meeting all delivery dates of customer orders* (the delivery date is greatly significant compared with all other boundary conditions mentioned earlier and therefore is relevant for most production environments), the MES must ensure the following:

- Synchronization of the process chain by means of the parameters in the work plan to minimize processing time. This means, among other things, avoiding idle times and waiting times (e.g., minimizing storage costs for the production warehouse) while simultaneously considering resource requirements.
- Collision-free planning of an order pool in the respective time container, taking into account the specified priorities and rules for optimizing sequences.

5.4.4 Strategies for Sequence Planning and Planning Algorithms

In addition to the requirements shown thus far in this chapter, the algorithms used for *sequence planning* are the deciding factor. For simplification, we speak of an *algorithm* in this context, although the planning system may include a complex set of rules, a simulation system, or even an *expert system* with self-learning software components.

The use of *simulation tools* is essential for optimal planning. The reasons for this are obvious—simulation tools are already used in the factory-planning phase for coordinating machines, equipment, and logistics processes. As a result, the most important boundary conditions are mapped in these systems. The simulation tools for planning an order pool are different from those used in factory planning and product development. With the latter, the following parameters are emphasized: quantity, date, calendar, shift model, alternative machines, and variants.

The simplest variant of planning is the *interactive control station*. Here, planning is carried out on a classic planning board. The planning data transferred from the ERP system are imported and visualized graphically, and in the event of capacity overload of individual machines or equipment, *capacity equalization* is carried out through manually delaying orders. Thus no independent planning algorithm exists in the MES, but instead, the humans assume planning with all the advantages and disadvantages this entails. In particular, planning for preliminary products from in-house production mentioned earlier is difficult here because the entire process chain (and parts list) is not resolved by the MES.

In order to avoid this "emergency solution through manual delay," a *planning algorithm* must be able to resolve and synchronize complex process chains and to carry out collision-free planning of a time container with a large number of orders, taking resource availability into account. Changes to quantities, dates, or shift models are entered manually. The algorithm determines the rest. The planning result then is displayed, for example, as a *Gantt diagram*. This planning then is contrasted in turn with actual production data.

Gantt Diagram

A Gantt diagram, or bar chart, is an instrument of project management named after the analyst Henry L. Gantt (1861–1919) and represents the chronological order of activities in a graph in the form of bars on a time axis.

In contrast to a network plan, the duration of the activities in a Gantt diagram is clearly visible. One disadvantage of the Gantt diagram is that the dependencies between activities can be displayed only in a limited manner. This, in turn, is the strength of the network plan. [SYSKA 2006]

5.4.5 Forward Planning/Reverse Planning/Bottleneck Planning

For the optimization-based planning of orders, there are different strategies that are described briefly here. The appropriate strategy should be selected depending on the initial situation and boundary conditions (Fig. 5.3). The individual production orders consist of subassemblies A1 and A2 and B1 and B2, respectively. The hatched areas represent setup processes of the respective production steps.

Normally, the strategy of reverse planning is used. If the process chain for fulfilling an order extends into the past, a planning system must switch automatically to forward planning.

In *forward planning*, the MES is provided with the earliest possible production start based on the material resource planning (MRP) run (verification of material availability). Planning is carried out based on this date and beginning with the lowest production level (secondary requirements). All necessary production steps are scheduled moving forward in time. However, this is not always the suitable method. If the final product is finished too soon, higher warehousing costs may arise owing to the added value of the end product, and it is also possible that raw materials used could have been better directed to an urgent order.

If the customer has been given a delivery date, or if it is possible to deliver only on certain dates, for example, because of shipping schedules, *reverse planning* of the production order is recommended. In this case, planning is done based on the finish date of the production order with the highest production level (primary requirements). Here, the individual production steps are scheduled using reverse planning.

Often, production steps for different products arise that all must be carried out on a particular machine. In this case, *bottleneck planning* is

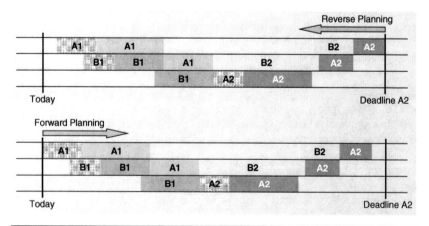

FIGURE 5.3 Reverse and forward planning.

carried out based on the bottleneck resource. The production steps are scheduled using a combination of forward and/or reverse planning.

5.4.6 Collision-Free Planning of a Time Container

In the production industry, planning an individual order tends to be the exception. Generally, there are a large number of orders that need to be fulfilled "simultaneously" as per the requested dates using limited resources and capacities.

Here, the aim is to find or calculate an optimum for the sequence, cycle times, and storage costs. The calculation algorithm must be capable of planning the sequence according to priorities and rules without collisions and with minimal gaps.

Collision-free calculation with a manually established sequence of individual orders requires more effort from the planning algorithm. Every order consists of various operations, and the MES therefore must schedule a large number of operations. Then a possible delivery date can be stated for every customer order. The speed of the calculation depends largely on the number of operations to be planned in the time container. For this reason, it is advisable to consider the shortest possible periods of time. If the planning period of a time container has expired, open orders must be listed in order to move these into the next time container.

In accordance with the parameter settings made and the sequence method, an algorithm determines collision-free planning with minimal gaps and exact delivery dates. In the example, 10 orders for the same article with a quantity of 100 pieces and the availability date 20/12/2007 at 14:00 are placed in the time container. The planning algorithm calculates the individual delivery dates of the 10 orders based on the planning strategy selected; these are provided as an overview (see Fig. 5.4).

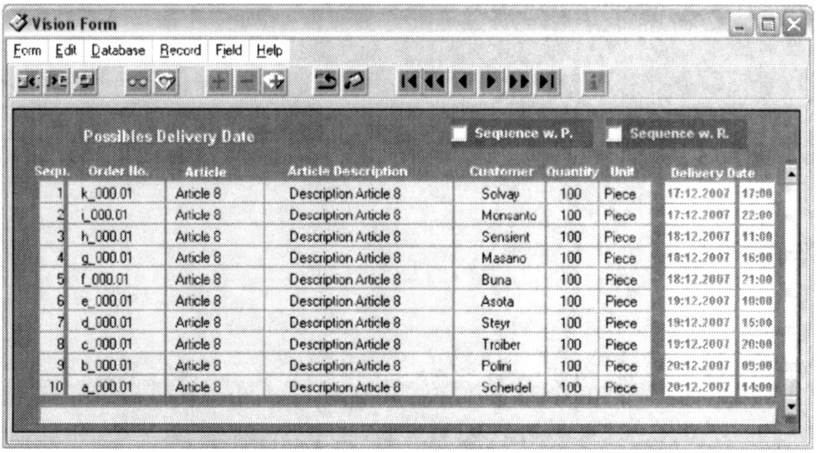

FIGURE 5.4 Overview of an order pool.

5.4.7 Setup Optimization and Warehousing Costs

The result of the planning also should include an exact determination of planning costs that takes not only the calculation of direct costs but also the allocation of overhead and warehousing costs related to the product into consideration. This calculation should show how a sequence optimization affects setup costs and, in parallel, storage costs. A complete MES should include this function. For example, it could be that the savings achieved through setup optimization increase the warehousing costs to such an extent that setup optimization is not practical. Such and similar situations should be clarified by the MES in order to show the responsible parties the most economical alternatives.

5.5 The Importance of the Control Station

5.5.1 Core Elements

The terms *control station* and *control room* was coined in the initial phase of computer-supported automatic production and are still used in various contexts. In process technology, they mean display of the entire technical process in the form of flowcharts. In the scope of an MES, a *control station* denotes the user interface for order planning. This means that the result of automatic planning runs is visualized, and/or an interactive planning or correction of the suggested planning can be carried out by humans.

One core element is the depiction of the planned orders as a Gantt diagram, often also referred to as an *electronic planning board*. If the MES has planning algorithms for optimization that enable it to make collision-free calculations of even complex process chains with any number of orders, the graphic display is essential. In this case, only a little interaction is needed, which leads to a simplification of the user interface and therefore helps to avoid errors. Typical algorithms help to optimize

- Setup costs
- Production costs
- Personnel costs
- Capital lockup
- Adherence to deadlines
- Capacity utilization

In very few cases, production will use only one optimization strategy (e.g., meeting deadlines at any cost). Usually, optimization arises from a weighted combination of the algorithms described.

If planning data are simply imported from the ERP system, for example, because of a missing planning algorithm, the control panel must clearly display collisions and multiple entries of machines or equipment (three-dimensional load diagrams). Reducing load conflicts and thus equalizing machine utilization is done by interactively postponing orders. This function must be made available in the simplest, most transparent form possible by the control station. Intuitive usage is very important to avoid misplanning. In Sec. 5.5.2, the main content is described in the form of a sample structure of a control panel.

5.5.2 User Interface

Overall display should be simple and uncomplicated based on the guidelines for user ergonomics. The less space wasted on information that is not absolutely necessary, the more is left for the actual core information—the graphic representation in the form of a Gantt diagram. Since the requirements in different sectors and production areas can be hugely different, no general statement as to which data are especially important and when and for whom they are possible. Therefore, the user-friendliest solution is an interface configuration that is as free as possible. The system user configures the control station for his or her own needs and saves these settings. When the user next logs into the system (ideally, independently of the workstation at which the user logs in, e.g., using a Web client), he or she receives his or her customized user interface. The example in Fig. 5.5 was structured into five normally required sections.

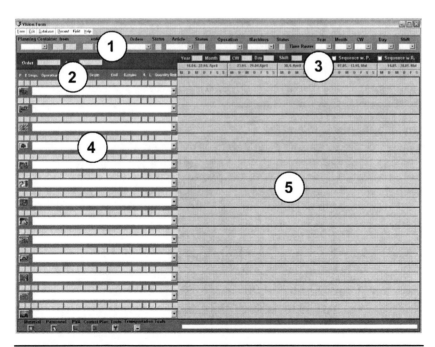

FIGURE 5.5 Basic structure of a control station.

Information Filtering Section (1)
It should be possible to select the planning period (a target value; e.g., the current calendar week should be adjustable) and the planning object (e.g., machine, station, article, etc.). The selection made must be clearly displayed.

To use this filtering function efficiently, great value should be placed on the arrangement of the recorded data in tables. This arrangement should be based on the methods of a multidimensional table structure using online analytical processing (OLAP) tools.

Selection of Planning Object (2)
The display can be made either via production orders (machine view—selectable from the actual order or the article produced) or machines/lines/stations (order view). In an order view, the individual resources are displayed on the y axis (4), and the operations of the orders are displayed as time bars in a Gantt plan (5). Therefore, it is possible to see with which orders a resource is occupied and how the resource is used. In machine view, the selected orders are displayed on the y axis (4), and the operation-resource combination is displayed in the Gantt plan (5). For example, it is then possible to display the operation of a production order via the machines/stations.

Configuration of the Time Axis (3)
The time axis can be configured to suit the selected time period and time pattern. Specification arises automatically from the time period selected.

Data Display Section (4)
Lines/machines/stations or orders corresponding to the selected planning item are displayed (2). To clarify the display, images of the machines or article can be faded in.

Graphic Display Section (5)
Lines/machines/stations with orders or the orders with their operations as time bars corresponding to the selected planning item are represented (2). This display of the planning data can be contrasted with the actual data recorded by the MES. Optionally, the actual data can be imported immediately into modified planning.

5.6 Personnel Planning and Release of Orders

The personnel calendar is important for production management for more long-term planning of personnel utilization. This is updated in the MES and linked with the company's time model (shift calendar).

In the short term, that is, shortly before releasing orders for the execution of production, it is important for the shift manager to know whether the required personnel with corresponding qualifications are available. Only then can an order be released for execution. Therefore, the attendance situation and personnel qualifications must be checked in this planning phase. The shift manager must constantly monitor the personnel capacity that is still available in the current shift when scheduling personnel (see Sec. 4.3.2). Every person can be allocated to a separate calendar. Within the calendar, individual shift models can be used.

5.7 Summary

A qualified MES has an operative planning system that carries out collision-free, situation-based planning of an order pool, taking the resource situation into account. The system ensures that order execution is subject to a realistic time frame and that possible deviations are recognized immediately through ongoing comparison of target/actual values.

Fundamentally, such a system must include planning algorithms for optimization based on realistic work plan data. The result of planning calculations can be displayed as a Gantt diagram for a better overview. Interactive control stations usually only hide the lack of qualified algorithms.

Planning calculations take the customer view into account by orienting the sequence of orders according to the priority, the weighting of the respective customer, or internal cost minimization. This affects the application of technological rules (e.g., setup matrix). The effects on warehousing times and the associated costs must be taken into account. In some cases, savings from optimizing setup may be offset by additional storage costs for the production.

CHAPTER 6

Core Function— Order Processing

6.1 General Information on Order Processing

6.1.1 Classification within the Overall System

Unlike the core components mentioned in Chaps. 5 and 6, which are indirect added-value elements, order processing is closely linked to the direct added-value process (see Fig. 6.1). Here, it is important that the machine staff have a user interface [manufacturing execution system (MES) terminal] that can be adjusted to fit precisely the needs of the respective production processes, thus guiding the user simply in the sense of a workflow.

The successful implementation of an MES applies here. Problems with the implementation of information technology (IT) in the production-related sector are very often caused by the noninvolvement or delayed involvement of the machine staff in the collection and configuration of information (see Chap. 9). Involving the worker means that the worker should first be informed regarding the goals and meaning of the MES. In the second stage, the MES terminal should be developed and tested. This also includes equipping it with the necessary hardware, such as the size of the screen and the interaction medium (e.g., mouse, touch screen, keyboard with function buttons, staff identification unit, and interfaces to machines). Before expanding on the individual functions for order processing, a short overview is given and two possible variants of an MES terminal are described.

6.1.2 Functions of Order Fulfillment

The worker at the machine or manual workstation is responsible primarily for processing an order. Therefore, the main focus is on the necessary functions for this task. These are described below. In automated processes, an MES must exchange the information necessary

100   Chapter Six

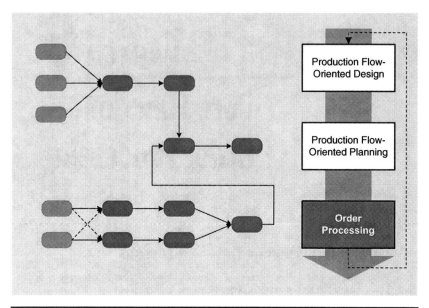

FIGURE 6.1   Core element: Order processing.

for order processing directly with the control systems. MES terminals are not necessary here.

Departments such as production management or materials management are also involved in processing an order because only if sufficient material is available at the machine at the right time is the worker in a position to fulfill the order according to target. However, controlling and company management are also involved in order processing. The MES also must provide formatted information to these departments at the right time to ensure that decisions can be made.

6.1.3   The MES Terminal

The demands a terminal has to meet with respect to recording and information functions depend on the company and the sector. For production, when it comes to information systems, we differentiate mainly between whether an employee uses a PC (with a mouse and a keyboard) or an industrial PC (IPC) with a touch panel as an MES terminal. Both system variants have their pros and cons. The PC generally has a larger screen so that more information can be displayed at the same time. In addition, information can be entered and selected more quickly using a keyboard and a mouse. However, these systems are not particularly robust and therefore may be unsuitable because of factors in the working environment (e.g., vibrations, temperature range, dust, etc.). Here, compact models with touch panels can be used. Workers can use these directly from the screen—in extreme cases, even while wearing gloves.

Core Function—Order Processing 101

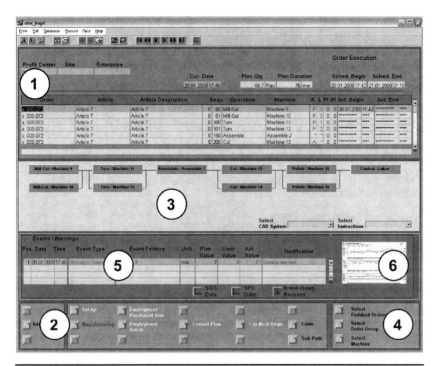

FIGURE 6.2 Example of a PC-based MES terminal in production.

The structure described below has been designed for a maximum of required content. In individual cases, the content may be reduced depending on the application. The PC-based variant is presented first in Fig. 6.2.

List of Orders (1)
In the first frame, the orders allocated by the shift manager to each machine worker should be displayed in the sequence in which they are to be processed. Allocated to the order are planned quantity, planned duration, possible planned idle time, scheduled begin, and scheduled end. When the order is started by the worker (with respect to the current operation), the actual begin is determined.

Capture and Control of Order Data (2)
Workflow control begins with the commencement of the order in the selected operation. The MES then should offer the *operating data acquisition* (ODA) functions. The ODA functions can be divided into the following subfunctions: setup, production, and disassembly/cleaning. Each of these functions is activated only if it is needed for the current operation. Within the individual functions, further subfunctions can be controlled. These individual functions also can be allocated to the *production data acquisition* (PDA) main functions. In the example in

the figure, most of the functions are allocated to the main function "production." Here, the machine user should be guided on the terminal in the sense of a workflow—only those individual functions are activated which are to be completed in the current operation.

Graphic Representation of the Work Plan (3)
In order to give the machine user an overview of the work plan, this should be displayed optionally in a graphic diagram.

Calling Up Information (4)
Optionally, it should be possible to call up information on finished orders, machine performance, and the status of already completed work processes.

Displaying Events and Exceeding Limits (5)
If events such as machine downtime or the exceeding of limits (e.g., 6Sigma violation) arise, these should be displayed for the user in real time with a link to measurement data.

Display of SPC/SQC Data (6)
This frame is intended to display the results of the statistical process control (SPC)/statistical quality control (SQC) function in diagrams.

As described earlier, however, this variant is not suitable for all areas of application. One possibility already mentioned is the use of an IPC with a touch screen at the terminal (Fig. 6.3).

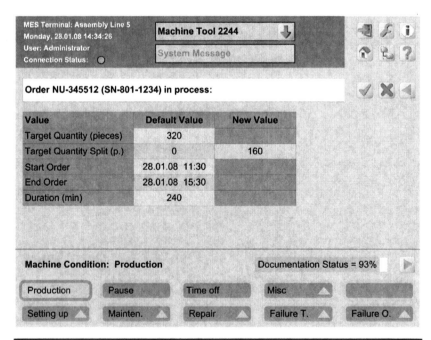

FIGURE 6.3 Sample illustration of an MES terminal as an IPC with a touch panel.

The information contained in this figure is equivalent to the PC-based variant described previously (see Fig. 6.2). The size of the screen means that it is not possible to display all information to the machine operator at the same time. Furthermore, large buttons and fonts are needed to make it possible to use the IPC wearing gloves.

In the standard view, the order list and the buttons for selecting the machine condition (e.g., production, break, maintenance, etc.) should be visible. The user should be able to display all further functions optionally. For complex functions, such as process instructions, test instructions, or assembly information, the worker can be guided through the workflow by "wizards."

If additional detailed information is required on an order, this can be displayed by selecting an entry from the list. When an order has been selected from the list, it can be launched using ENTER key. Alternatively, it is also possible to launch an order via a hand scanner, etc. by reading a bar code.

As will be demonstrated in the next section, it is especially important for subsequent automatic evaluations by the MES that the machine condition (e.g., production, maintenance, error, etc.) is fully documented. Only thus can it be guaranteed that the core figures calculated are meaningful. This condition is recorded automatically using sensors to the extent that this is technically possible. However, the machine also can be in a condition that is undefined from the machine's point of view. In this case, the worker must select a cause of disruption from a list. This so-called after-documentation also can be made after the end of the shift for all undocumented conditions.

If the machine is in production, the quantity produced must be recorded. This information also should be recorded automatically, if technically feasible. Manual input or possibly a combination of both manual and automatic input also should be possible.

6.2 Order Preparation and Setup

6.2.1 Changing Tools

As a resource, tools should be made available and installed product-specifically. When the function is triggered, the beginning of use of the tool is recorded on the machine. In addition, the status of the tool is reset in the respective data set according to the situation.

Installing tools is often permissible only when complex regulations are adhered to. Therefore, it is necessary to display the relevant regulations for the worker on the terminal during installation. Depending on the size of the machine, a mobile display device also may be useful for personnel.

The disassembly process proceeds in a similar manner. Here, the tool is released once more, booked into the warehouse, and the accumulated usage time is added to previous usage time. If the accumulated usage time exceeds an established limit, this must trigger an order for preventative maintenance (see Sec. 6.5.2) or, at least, a warning message must appear.

6.2.2 Machine Settings

A decisive factor for an integrated MES is an online connection to existing machines. Modern technologies make it easy to establish interactive communication between the MES (level 3) and the control level (levels 0, 1, and 2) (see Sec. 7.3). This communication is bidirectional. This means that process-relevant data can be loaded into the machine in an order-based fashion, and machine status also can be forwarded automatically from the system to the MES. Thus, when this function is triggered, control programs [e.g., distributed numerical control (DNC) in discrete production and workflow processes in the processing industry] are loaded to the machine controls according to the definition in the product master data and are activated there. It is still possible to begin recording, visualizing, and controlling data for the machine parameters. It is also possible to connect several machines to one terminal (see Fig. 6.4).

Supervisory control and data acquisition (SCADA) systems that visualize data in real time and assume functions of the MES terminal also can be used as a link to the MES (see Sec. 3.4.4).

Since in some cases machine settings must be adjusted manually, for this function it is also necessary to display the setting parameters

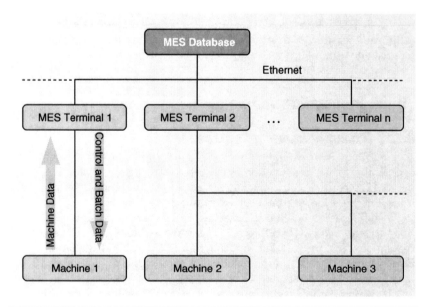

FIGURE 6.4 Coupling the MES with the automation level.

and possibly also the related setting instructions to the machine user on the screen. These also include instructions for cleaning processes, which are to be carried out in an electronically executed workflow as per the guidelines of FDA 21 CFR Part 11 (see Sec. 3.2.4).

When machine settings have been completed, the production function can be started at the terminal.

6.2.3 Material Provision

The material provision can be carried out within setup, and the required provision time can be set. If the provision process is associated with greater outlay, a separate order with all possible partial functions is generated within this operation in order planning. The material provision list must be issued for personnel when this function is triggered. It is important that the material can be identified easily according to warehouse organization. A plausibility test must be carried out for the material to be provided on the card accompanying the material, the warehouse position, and the related withdrawal posting. If feasible, all material for the ordered is made available.

The situation is different if a *kanban* system is used within the MES; that is, the material is requested according to the production situation on the basis of *kanban* settings. Basically, the *kanban* method provides that when the order is started on an operation, only a certain starting quantity of material is requested initially. The material then can be reordered repeatedly.

For example, at the beginning of the early shift, there is warehouse capacity for 20 pieces for a production lot of 20 articles. As a starting quantity, 20 pieces is entered in the master data. If 10 articles are produced with 10 of these parts, a *kanban* provision order is issued requesting delivery of another 10 pieces of the material. The required control parameters should be managed for each product in a *kanban* operation.

6.2.4 Test Run

When the machine has been retrofitted with the suitable tool and the correct settings have been made, a test run can follow, provided that the material is available. This is important because this process ensures that the finished product corresponds to the specifications. The production tolerances, etc. should be tested. It may be necessary to make adjustments to the numerical control (NC) program or optimize recipes.

If the worker is not able to determine the production tolerance/quality, samples must be sent to a measuring station or a laboratory. For this purpose, the MES terminal should provide a relevant function that allows for notification of the measuring station/laboratory. The machine user can begin production of the order only after the sample is approved.

6.3 Order Control

6.3.1 Information Management

The provision of information is essential for order control. First and foremost is selection of the correct order. This is provided to the worker directly on the machine via an MES terminal as a list of pending orders (see Sec. 6.1.3).

However, process or testing instructions also must be provided through the MES. This affects, among other things, the maintenance technician who requires the appropriate machine documents, guidelines, etc. on-site for his or her work. The documents are managed by the MES as document management.

Irrespective of the terminals available, it is useful to provide workers with additional special information, such as target and actual quantities of pieces for a running shift. This can be done practically by displaying this status information from the respective production areas on large displays (and on boards; see Sec. 7.4.1).

6.3.2 Control and Tracing of Production Units

In order to control the production units (e.g., piece, component, batch, or lot), it is necessary to trace them (tracking), that is, to know their exact whereabouts, status, etc. at any time. The term *tracing* describes how the exact production process can be reconstructed ex post with all important events and data.

All nonidentifiable parts/materials (e.g., screws, bearings, or stickers) are allocated to what is known as a *pool*. It is possible to record all relevant details such as disruptions, messages, process values, attributes, parameters, and measurements up to a list of parts used via the MES. This collected data are archived and are, as described, visible ex post.

6.3.3 Managing the Production Bin

The goal of a qualified MES is to ensure a continual production flow. In practice, this is possible only under ideal conditions, which seldom exist. In reality, buffers (production bins) arise between the individual operations. These are associated with warehouse bonding times and, consequently, with warehouse bonding costs. These times and costs are to be retained in the MES. However, by using an MES, it is possible to reduce this production bin to a minimum.

When the last operation for an article is reached in order processing, the produced quantity is booked to its final storage place. The warehousing times and costs are allocated to the end article.

6.3.4 Material Flow Control

Within the execution process of an MES, coupling material usage functions with the creation of the individual outputs (in the operations) is key.

Control of the preceding and succeeding relationships in the process chain is made possible by the MES.

Material Usage Function
The goal is to provide the required material in the correct quantity and quality to the workplace at the planned time. The value-creation process can begin only when the required material is available at the correct location. This occurs analogue to the provision process using a material usage list; the material provision list possibly also can be used.

The material provision list and material usage list contain the material number, the delivery number, and the respective delivery position. This information is recorded using bar codes, for example. When the material is used, the bar code has to be scanned. Therefore, the status of the material should be changed. (from "provided" to "in production").

In the first operation, entire delivery units (e.g., a pallet) generally are delivered to the machine workplace with raw material, purchased parts, or preliminary products. If parts of this delivery unit are not needed, there must be an option to rebook the remainder back into stock in order to record the "remainder."

An MES should be flexible enough to ensure that additional quantities of the required material can be used from other deliveries. A special form of material usage is weighing recipe ingredients and using recipe materials in a sequence procedure.

Recording Output in an Operation
After the worker is identified, the container type and the input of the quantity are selected (if needed) if this information is not provided automatically via a quantity counter. If scrap and reworks are identifiable in the output, they should be recorded, and rework orders or additional production orders based on scrap must be made automatically.

Where quality tests are to be carried out for the operation, recording should be integrated at this point. This also applies to samples that must be taken for a laboratory order. Each sample taken goes to the laboratory with its own identification number.

In the MES, it must be possible to allocate any data from the process to any production unit. This can be a single piece (e.g., identifiable by a series number) or also a container with a certain quantity (Fig. 6.5). One of the goals of an MES is to ensure paperless production. The previously used accompanying product card is a document that is required for identification and specification of the next operation, etc. These data on the accompanying product card can be made virtually accessible to the person virtually on the terminal or as a paper printout as before. In both cases, the minimum data shown in Fig. 6.6 must be contained on the accompanying card for the product.

Chapter Six

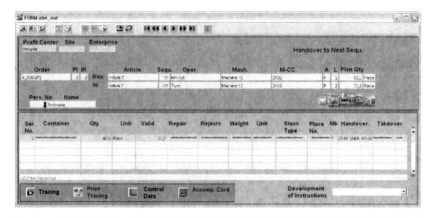

FIGURE 6.5 Output recording.

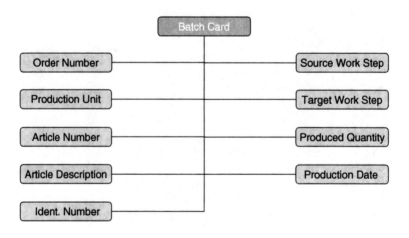

FIGURE 6.6 Minimum data for the accompanying product card.

The much-discussed topic today of material flow control using radio frequency identification (RFID) technology is highly interesting from a technological point of view. In practice, however, accompanying documents are often still used in paper form. This variant definitely has a valid reason for existence in modern production processes because it is sometimes simpler to manage the process this way and be able to do without additional terminals, handheld devices, etc. (for displaying information on the virtual accompanying product card).

A standardized MES also must be proficient in the following special forms of output recording:

- *Reel production and reel transformation.* In this variant, for example, the focus is on transforming mixed approaches (e.g., quantity unit liters or kilograms) into discrete units (e.g., reels with quantity unit meters) in the manufacture of synthetic materials and generating several transverse and longitudinal reels with a unique identification number from single reels.
- *Mixture of several batches used.* With liquids, the mixture percentages of the charges used should be determined in the output (e.g., mixture of various decoction batches in the brewery). This must be done dynamically for refilling processes.
- *By-products.* If by-products arise in an operation, they must be managed by the MES and must be processed further in their own process chains (e.g., mash in the brewing process).
- *Output for further processing in several articles.* In an MES, it must be possible to divide the output of an operation or production unit into several articles. Each of these new articles can have its own process chain. These articles can flow once more into the original article in a subsequent operation.

6.3.5 Order Processing and Operating Data Recording

Operating data acquisition (ODA) is the function group for recording and controlling all production performance data and includes both the preparatory measures (i.e., setup, dismantling, and cleaning functions) and the direct value-creation functions. Setup, cleaning, and dismantling processes are usually necessary for every machine. They often take up a large portion of the cycle time in a single operation. As a result, "optimal" sequences are determined for an order pool via optimization algorithms in sequence planning (see Sec. 5.4).

In the ODA "setup" function, the individual machine is prepared for the actual production process. These are processes such as

- Material provision
- Tool integration
- Setting machine parameters
- Startup activities
- Cleaning processes

Performance recording for these activities is quantity-dependent. When the function is started, starting time is recorded. This generally applies to the order and to the personnel who have been allocated by the shift manager in personnel resource planning. It always must be possible to deploy additional staff where needed.

The direct value-creation process function "production" is the classic field of ODA, with the task of documenting who, when, where, what, how long, and how much is produced. This concerns recording and controlling the performance process, which is recorded and controlled in a quantity- and time-specific manner. From this function, it is also possible to branch into other functions, if needed (e.g., owing to interruptions of the production process or for information functions):

- Processing of downtime (e.g., duration of downtime, reasons for downtime) with the activation of maintenance control (e.g., ad hoc maintenance, preventative maintenance)
- Material flow control, logistics process
- Quality assurance in the process and quality control for the product (SPC/SQC)
- Handling of scrap and rework
- Performance analysis, including cost control
- Traceability

In individual cases, it should be determined to what extent these subfunctions should be integrated into the setup or production functions. The times for the staff used and the machines normally are already documented in the setup and production functions.

Either the necessity for a rework is recognized directly in the process, and a rework order is created, or this is done by the subsequent quality assurance. Here, it is also decided whether partial quantities are declared scrap and whether a corresponding rework or remanufacturing order should be processed. It is important that these unforeseen orders are taken into account in the order planning. In addition, it also should be apparent to which planned order the rework or remanufacturing order belongs. The unforeseen orders then are treated as planned orders—with all recording and controlling functions.

6.3.6 Process and Quality Assurance

Overview

After an initial boom in quality assurance systems at the end of the 1980s, these systems are experiencing a renaissance today—in part with new concepts. However, significant here is the integration of quality management in production management. The control of quality can be subdivided into the process control of the machines with the statistical process control (SPC) functionality and the actual quality control of the products created with statistical quality control (SQC) methods.

In the end, it is always about keeping the production process stable through systematic methods so that the maintenance of defined control parameters is statistically ensured. The limits for these parameters are either provided by the customer or defined through internally provided action and warning limits. The statistical equipment must be used in such a way that, at least for process supervision, the statistical controls run automatically in the background, and the SPC system makes decisions itself in the case of deviations that are statistically unacceptable (e.g., the scrapping of parts).

An integrated quality management system is a core function of the MES. It is the task of the MES to carry out continuous checks. Thus it is the basis for such concepts as define, measure, analyze, improve, and control (DMAIC) and 6Sigma (see Chap. 8).

Process Supervision
The targets for machine control, both for discrete and process-oriented processes, arise in the course of setup. During production, the recording and control of the parameter data occur; this also can be multilevel.

By means of process supervision, tests should be carried out continually to check whether the measurement values are *capable of being processed* based on the target sizes of the control parameters (e.g., target average, lower and upper target limits). As well as the integration of the statistical equipment, this requires a well-thought-out data structure both on the real-time level of the machines and on the MES level in which the measurement data are retained and analyzed in a compromised format. With the data structure, it should be taken into account that saved arithmetic averages also can be summarized once more into new average values for longer periods of time (i.e., aggregation).

In the MES, the data for the measurement cycle also should be managed as well as the testing parameters. Here, we mean the measurement lot sizes according to which the short-term *process capability index* (C_{PK} index) is determined or checked if a significant change in spread, situation, and distribution has arisen. The long-term analysis of process capability (see Fig. 6.7) observes a large number of measurement lots (e.g., charges) and reproduces the entire performance,

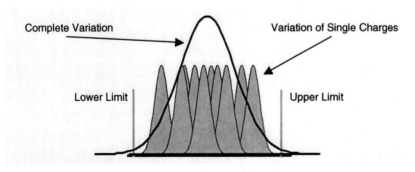

FIGURE 6.7 P_{PK} index for the representation of the overall performance.

including "drift" among the charges. The *process performance index* (P_{PK} index) should be at least greater than 1.67.

Quality Control
Even if a qualified SPC system is in use as part of the MES, the finished products must be made subject to quality control on the basis of random sampling plans. In the sense of a dispatching function, random samples (e.g., third, fifth, etc., random samples or dynamic random sample sizes) should be taken as per a testing frequency established in the testing plan (e.g., every hour). This should be realized within the scope of an integrated SQC system. The process parameters of the process supervision are the influencing sizes, which are generally measured and recorded manually by SQC. For SQC, the data are evaluated statistically and made transparent in graphic form by means of what are known as control cards (e.g., xcross/s card), histograms, etc. Here, it must be possible to carry out recording specially for each production unit both for variable and attributive characteristics. For attributive characteristics, error types should be allocated and then visualized in error-collection cards.

If limit values are exceeded, the SQC system can initiate the execution of measures that are to be executed and acknowledged electronically as per the guidelines of the DIN EN ISO 9001 and FDA 21 CFR Part 11 standards (see Sec. 3.2.4) in a workflow.

6.4 Performance Data

6.4.1 Involved Departments

The following sections of a company are affected by events and warnings from production:

- Production management
- Control
- Sales and marketing
- Process control
- Quality assurance
- Maintenance
- Material management
- Logistics

Production Management
Production management has the central control function and is responsible for all performance functions of the production. All deviations from specified parameters are of interest here. Therefore, the production manager needs an information and control view with a real

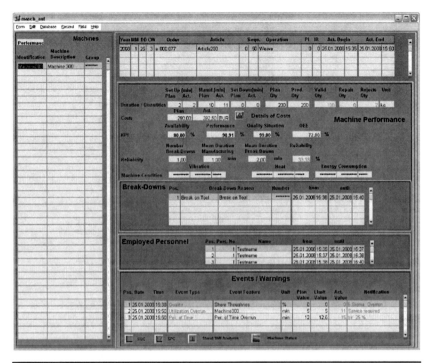

FIGURE 6.8 Information panel for production performance.

picture of the production. Any unacceptable deviation from the target must be visualized in order for the manager to be able to implement the necessary measures and inform management in the case of larger problems.

Here, an information panel (also known as a *dashboard*) is suitable, and this is described in Fig. 6.8.

From this information panel, it is possible to branch into a schematic display of the machine infrastructure from which the status of the facility can be read at a glance. By clicking the individual sections, detailed views can be displayed (see Fig. 6.9).

Furthermore, the individual, production-relevant events should be displayed (see Fig. 6.8).

It is possible to distinguish the following fundamental event types with various control characteristics:

- Process reports (measurement parameters)
- Quality reports (test parameters)
- Maintenance reports (exceeding of usage limits; the maintenance time results from this)
- Material stock reports (safety stock infraction)

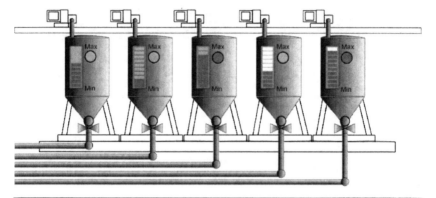

FIGURE 6.9 Schematic representation of silos at a control center.

- Date control reports (exceeding time limits)
- Cost control reports (exceeding costs)

The events usually are reported by individual machines/systems. These can be filtered and grouped by the MES so that already preprocessed reports can be made available to workers. The data set shown in Fig. 6.10 should be retained for any event of the defined event types.

There are events that only generate a report when they arise. This report then can be acknowledged in conclusion. However, it can be entirely useful to create an additional report with the "going" of the event. For example, it makes no sense to be able to acknowledge an alarm report for an overheated reactor container as long as the container is still outside the permissible temperature range. Here, an

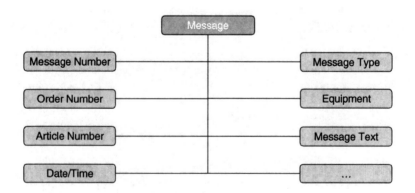

FIGURE 6.10 Components of an event report.

MES offers various acknowledgment mechanisms that are linked with various report classes.

The production manager receives all information that is relevant for him or her in an overview. He or she can look at the data received more closely, such as the process of individual process characteristics of a machine. The person responsible for a particular control area has, depending on the information philosophy, only those events displayed that affect him or her directly. It should at least be ensured that he or she exclusively acknowledges reports from his or her area.

Business Management
Business management is interested primarily in whether the planned costs for the orders and the delivery dates are adhered to. In the future, business management will be included in performance process supervision with selected information in the sense of an early warning system.

Control
Control has a focus on cost control in real time. Here, only the cost deviations are displayed, and the controller has the option to have the reasons for deviations displayed.

Sales and Marketing
Sales and marketing receive date statements from the MES, which can be achieved under pressure from operative order planning and also during execution of the production orders. With these, the sales and marketing worker can pass on reliable delivery dates to the customer at any time. In order to ensure that the dates are also adhered to (e.g., in the case of customer queries), sales and marketing also fundamentally need relevant event reports, which affect order progress.

Material Management
Material management fundamentally receives warning messages if additional material requirements are reported in the course of operative order planning or if security stocks are fallen short of.

Process Control
For process control, unstable processes or critical machine situations are of interest. If limits are exceeded, this is reported, and the person responsible for the process can view the event types in order to take corresponding measures.

Quality Assurance
When the limits for process and testing parameters are exceeded, the events are reported to quality assurance. All events are displayed with the event types "process" and "quality."

Maintenance

The maintenance department receives maintenance reports and is notified in cases where use limit figures are exceeded. The due dates of existing maintenance work for machines and resources are visualized in a *maintenance plan*. Furthermore, the department receives reports about disruptions to machines or facilities and downtimes, as well as the implementation of disruption resolution (ad hoc maintenance).

Logistics

All reports about the material flow, warehouse bonding costs, and the related warehouse bonding costs are forwarded to the logistics department.

6.4.2 Key Figures and Performance Record

In addition to order-based analysis, it is also important to record the performance of the machines/facilities for defined periods and constantly monitor deviations. Subsequently, typical key figures are entered for the assessment of machine performance (KPI = key performance indicator) (see also Sec. 9.3.5):

- Availability
- Degree of quality
- Overall equipment effectiveness (OEE)
- Degree of performance
- Productive time, etc.

The guidelines for production, which have arisen in the past two decades, are aimed at ensuring that every company must document its performance process in order to make it possible to trace individual orders or charges consistently. Here, too, the MES forms the basis with its extensive recording processes. On the one hand, it is about being able to call up the production flow forward and backward at the touch of a button. On the other, the system must be able to archive defined process values, for example, recordings on environmental protection, for up to 15 years. Simple access to the archived values must be possible via a filter.

6.4.3 Ongoing Analysis and Evaluations

The core elements of the MES with which we have dealt until now guarantee the controlled running of the production process for the individual products and the documentation of this process with all relevant data. Performance control and analysis set extensive challenges for integrated information management. For the MES, three categories of performance control information are relevant:

- Event management, real-time performance control
- Performance analysis for products and resources
- Proof of performance for compliance standards

For every category, a set of standardized evaluations is made available both on the screen (at best as a Web client) and for printed evaluations (reports). In order to ensure that such an information system works efficiently, all relevant performance data should be recorded for the related target values. Lately, only deviations from the target situations are of interest. This means that an inundation of information can be avoided. The approach of accumulating data sets for defined time reports also plays a sufficient role here. This helps to reduce searching and calculating outlay.

6.4.4 More Long-Term Analyses and Evaluations

Overview
With more long-term performance analysis, the focus is on recognizing weak points and optimization potential in the production of individual products or individual resources (mainly machines/facilities). It must be possible to limit the analyses needed for this with flexible time filters, at least with the options of year, month, day, shift (each with a selectable value), and freely definable periods. For an article or resource (or a group of articles/resources), it must be possible to compare any time periods with each other. Here, more long-term changes, or *trends*, are recognizable. It must be possible to make a comparison for a defined period of time between articles and resources. This means that, for example, the question of which processing center has had the best OEE for the last quarter can be answered.

It also must be possible to evaluate the statistical data. For example, it can be checked how a particular process parameter for an article behaves on a machine with regard to the distribution form, or parameter ranges can be correlated by means of regression in order to establish dependencies. For this, there are a large number of tools that range from simple graphic evaluation options to sophisticated online analytical processing (OLAP) tools, including multivariate statistics (see Sec. 2.4.3).

Order Performance
With the analysis of order performance, the cost analysis generally is central. The arising costs are controlled in real time per work process and are displayed for early intervention. After completion of the order, these data should be provided to control in all details for conclusions and continuing measures.

Resource Performance with Regard to the Article
Important connections potentially can arise from the links between resource performance and the articles produced. Thus, for example, it can be recognized that the machine availability for a particular article is always very bad or that another article more often causes a tool breakdown. In this way, it is possible to analyze the causes of deviations in performance more closely.

Dependency Analyses
In the future, the methods of multivariate statistics (see Sec. 2.4.3) are to be implemented more strongly. This makes it possible to record dependencies between process parameters, especially through regression analyses.

Newer methods such as *part average analysis* (PAA), largely applied in the electronics and automotive industries, also should be mentioned. With this analysis method, each part is checked with a large number of parameters. If the sum of the measurements exceeds a defined limit, the component is declared ahead of time to be a defective part, and an automatic process locking may follow. Through continual process supervision, scrapping and reworking can be avoided or at least minimized. This means that zero-defect production is aimed for (see Sec. 8.2.2).

6.5 Maintenance Management

6.5.1 Tasks
Maintenance management integrated in the MES has the task of forecasting maintenance and repair measures and taking relevant preventive measures. A well-organized maintenance management within the MES contributes considerably to the achievement of quality targets.

Since disruptions within a highly automated production cannot be eliminated entirely despite preventive measures, a sophisticated alarm management system with short response times is also needed for the achievement of high availability values for the machines and facilities. This *ad hoc maintenance* also must be supported by the MES.

6.5.2 Preventive Maintenance and Repair

Preventive Maintenance Based on Machine Conditions
Today's maintenance management (TPM = total productive maintenance) is increasingly going the way of predictive maintenance, where maintenance is triggered for the machine based on condition factors. Here, online factors such as vibration, energy consumption, heat production of systems, etc. are monitored. If permissible limits are exceeded, a maintenance order is released or a warning message is provided at the machine terminal.

Preventive Maintenance Based on Usage Factors

As already explained in the setup function (see Sec. 6.2.2), facilities are set up and dismantled for setup, and their usage times are recorded. The usage time is also recorded in the "production" function across order processing for the machines and systems. A standard for usage can, however, also be a counted measurement, for example, for the recording of work cycles of an element.

If predefined threshold values for the usage sizes are achieved, for example, a number of hours of usage or a piece counter, the MES can automatically generate maintenance orders for preventive maintenance or display a corresponding report on the terminal.

Maintenance plans should be defined for maintenance of the machines/systems and facilities. Maintenance plans are constructed in a similar manner to testing plans (see Sec. 4.3.5) and contain detailed targets for the activities to be carried out and, where necessary, instructions regarding required replacement parts and resources. Maintenance plans also contain attributive control sizes with the related evaluations. The worker can print out the maintenance plan in order to make it easier to work on-site at the machine/system. Alternatively, processing can be carried out on mobile devices such as personal digital assistants (PDAs) or handheld devices. Maintenance work that has been carried out is archived by the MES.

6.5.3 Alarm Management

If an unforeseen downtime arises that cannot be resolved by the machine staff, a maintenance order must be generated for the maintenance and repair department by the machine user within the MES. For fully automatic systems, alarm reports with regard to maintenance are created directly. Notification of the employees affected occurs by e-mail, Short Message Service (SMS), pager, or directly as an event report on the screen of the receiver (i.e., an alarm report at the control center). The processing of the order occurs directly on the PC or on the mobile receiver device of the worker.

In order to monitor the availability of the individual machines (determining of key figures), the downtimes are recorded with the related times and reasons for the downtime. Where there is an online connection of the machines, the recording system automatically recognizes the downtime. Only in exceptional cases, however, is the reason for the downtime recognized automatically (e.g., for textile machines, recognition of a broken needle). As a rule, the worker/maintenance technician must select one from a list of predefined downtime causes on the terminal and allocate it to the situation.

6.6 Summary

In order processing, various sections of the company are involved both directly and indirectly. Each of these sections is provided with a tool by the MES, which guides execution of the work and, as a result, the worker given a defined workflow.

In production, the machine user receives the orders that are to be processed via terminals (or directly via the machine/system). All necessary additional information can be viewed online. Furthermore, all relevant data regarding the machine/system is passed on for evaluation via relevant interfaces on the MES. Other production-related sections, such as logistics and maintenance and repair, also obtain the necessary information and work orders from the MES. Work instructions for the execution of maintenance and repair measures are managed via the MES like the processing and depositing of testing protocols.

The processed data are provided in a suitable format to the production manager, the control department, and business management for evaluation. These evaluations are the basis for continual improvement processes. Data access is companywide (possibly worldwide) by means of Web technologies.

CHAPTER 7
Technical Aspects

7.1 Software Architecture

7.1.1 Fundamental Variants

If you analyze the architecture of the different systems, which are temporarily in the market, you will find two architectural variants with fundamentally different approaches:

- *Application-centered systems.* Here, the application controls the booking function in the database and the business logic of the system. The database serves only as a performance-saving medium.
- *Database-centered systems.* With this approach, the database is not only a data memory but also the pivot of the entire system. A large part of the bookings and also parts of the business logic are handled through the database.

Application-centered approaches offer advantages for development by the use of high-level languages. Updates are also simpler because the data structures are less complex. The main disadvantage is that errors in the application logic endanger the consistency of the data—in case of doubt, extensive repairs to the database by the manufacturer become necessary. In addition, core themes of a database management system (DBMS), such as cache consistency and transaction management, must be integrated through the increased use of multithreading; these have matured for years in DBMS with regard to freedom from errors, scaling, and performance. In systems with large data volumes, performance disadvantages also can arise because optimizations for the use of special DBMS strengths are not possible or need to be implemented several times and therefore are left out.

In database-centered systems, all data-related operations also occur within the DBMS in its native programming language in the form of what are known as *stored procedures*. A disadvantage is that platform dependency is no longer provided at this point. Software development also becomes more difficult because generally a high-level language offers more options than the programming language of the DBMS. This disadvantage is amplified by various current initiatives to make high-level languages (e.g., Java or .Net) directly useful for the programming of stored procedures. Updates are more elaborate because tests for integrity are an elementary part of the update process. The great advantage of the database-centered approach is that transaction management and the safeguarding of data consistency are transferred by declaration to the DBMS. The existing abilities of the DBMS, such as the support of cluster solutions or distributed systems, can be used without large adjustments to the application. The *data model* is a part of the application; that is, access becomes simpler for external systems (e.g., reporting) and is also possible without the contribution of the manufacturer in cases of doubt. A considerably better performance can be made possible by focusing on the DBMS with its specific optimizing options. This is particularly important for a manufacturing execution system (MES), which is confronted with a considerably higher transaction volume than an enterprise resource planning (ERP) system.

The following conclusion can be drawn from this consideration: application-centered approaches are easily at an advantage in systems with relatively low transaction volumes as well as in the early development stadium. For large systems with high requirements for availability, data integrity, and performance, on the other hand, the database-centered approaches are at an advantage. The considerations in this chapter therefore are based largely on this approach.

7.1.2 Overview of Central Components

The software architecture suggested in Fig. 7.1 for an MES is also present in many systems currently on the market in this form or a similar form. This architecture suggestion serves as an example and basis for discussion in this section, along with the database-centered approach discussed in Sec. 7.1.1.

Figure 7.1 shows a server-based system, the core of which forms a relational *database*. In order to be able to handle larger data sets, the database can be divided into two parts—an *online database* and an *archive database* (see Sec. 7.3). This database also takes on the basic functions of data processing, such as the booking in of complex data into various table structures. Laborious processing functions such as the calculation of key figures are handled in an external module. In this module (e.g., in the form of an application server), the *business logic* and *administrative tasks* (e.g., authentication of the user and datacare jobs) of the MES are mapped out. Independent *interface adapters* and/or *software services* (see Sec. 7.1.4) are available as interfaces to neighboring information technology (IT) systems. A server component is also needed for the *user interface* and the *reporting module* embedded in it. In the case of a Web solution (see also Sec. 7.3.5), this is a Web or application server.

Application Server

An *application server* is a component that provides a framework and various services for the execution of applications. Here, the concept *server* does not necessarily denote an independent hardware system. An application server provides special services such as authentication or access to directory services and databases via defined interfaces to the applications as a run-time environment.

The concept *application server* has developed to become one of the most used concepts in IT. Other concepts that are used in this context are *middleware* and *three-tier architecture*. Software applications with a three-level architecture generally are classified in the presentation, business logic, and data management layers. Applications for the mapping out of the business logic are also referred to as *application servers* in today's usage. Because of application in this middle layer, it is also referred to as *middleware*.

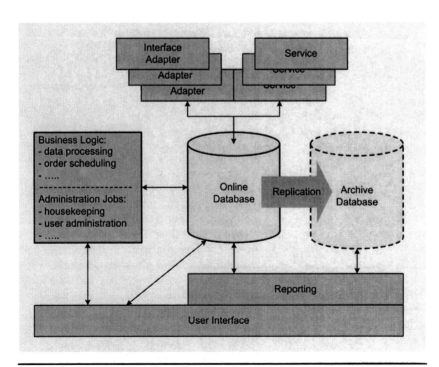

FIGURE 7.1 Central software components as an overview.

7.1.3 Platform Independence

The necessary outlay for the creation of true platform independence may seem high but must be taken into account with regard to the long system running times of an MES (generally more than 10 years). Changes in the IT landscape of the company may not lead to the end of the MES implemented; much more, platform independence should help to minimize costs for running and care.

At this point, the concept *platform* refers especially to the fundamental *computer architecture* (e.g., processors) *and the operating system used.* Naturally, the database also represents a platform, but this is not taken into account because there are only a small number of systems in the market that support different databases. Despite this, true platform independence is only given if the user also can select the database system freely.

On the subject of platform independence, two generally different viewpoints exist on the market:

- Users, generally in small- and medium-sized companies, who are not set on any particular platforms and therefore follow the recommendation of the supplier
- Users, often in large companies with independent IT departments, who have decided on internal conditions for platforms and also only permit systems that comply with these conditions

If a supplier wishes to serve both groups (so the entire potential market), this supplier's system absolutely must be *platform-independent*. In order to analyze this requirement more closely, the individual modules of the system (in accordance with the architecture suggestion from Sec. 7.1.2) are examined separately:

- *Database.* The database is the central core of the system. Therefore, the highest requirements with regard to scalability, availability, and performance are made of this module. Many suppliers support only one database system because it is hardly possible to carry out optimization for different databases. The database also requires regular software maintenance, for which the relevant skilled workers are needed. Therefore, if the database used is platform-independent (e.g., Oracle), an implementation on the platform preferred by the customer can succeed; if not, it is necessary to use a platform recommended by the supplier.
- *Business logic/administrative tasks.* Many of these data-processing tasks can be completed directly in the database using stored procedures. This method is distinguished by high processing speeds. The disadvantages are that more complex tasks can be mapped out only in Structured Query Language (SQL)

with difficulty and that change management is cumbersome. If you wish to avoid these disadvantages, you should use a Java- or C++-based approach. Both codes can be ported to different platforms with acceptable outlay (and so are platform independent to a limited extent), provided that you do not use any special libraries. From the viewpoint of user friendliness, the encapsulation into individual "functions" is desirable. For this purpose, the application of an *application server* (see "Application Server") or a *script engine* that contains individual jobs in the form of Java or JavaScript programs is possible.

- *Interface adapter.* Here, in turn, what we mentioned earlier under "Business logic/administrative tasks" applies to platform independence. It is possible that the handling of ports also can occur in the same software module. However, adaptiveness with regard to change is even more important for interfaces—no other topic is so often changed and requires as much time for implementation and testing. The use of scripts (at best, independent modules per interface) that can be changed simply and throughout the lifetime of the entire system is an adequate approach here. However, the possible existing infrastructure also must be taken into account, for example, if control systems of the production department can only be linked usefully via existing object linking and embedding (OLE) for process control (OPC) servers. A platform-linked solution (in this case, Windows) must be used here. OPC technology or Active-X-Controls are only available in Windows.

- *Software services.* The use of Web services has become established for a service-oriented architecture (see Sec. 7.1.6). The server-based implementation can occur platform independently in the scope of a Web application service.

- *User interface/reporting.* A Web-based approach is often also platform-independent. However, the many advantages of a "true" Web solution (using HTML and JavaScript exclusively) are still offset by the disadvantage of a sometimes low level of user friendliness.

7.1.4 Scalability

As is generally known, the only constant is change. If we follow this maxim, which applies somewhat for modern production, scalability is another important requirement of the system architecture of the MES in addition to platform independence. On one hand, the system must be adjusted as precisely as possible to the requirements of the customer, and on the other, changes in the production structure, that is, both changes in the scope of functions and in the quantity structure, must be easy to map out in the MES.

Changes in the scope of functions generally arise from the introduction of new products or from new ideas on organization and the changed activities linked with them. Examples here are the introduction of a worker information system or the switch to group work with a changed wage system. Many system suppliers cover such functional expansions with independent software modules, which are offered as add-ons for the basic software.

Scaling as per the quantity structure is more difficult and also affects the software architecture. The following key data of the system should be taken into consideration with regard to scaling:

- Number of machines and workplaces
- Number of articles produced
- Number of measurement values taken/sample rate for the measurement values taken
- Number and frequency of reports created (e.g., disruption reports from production controls)
- Number of simultaneous users (i.e., client stations on the network)
- Number and calculation cycle of key performance indicators (KPIs) and quality data
- Archiving period for KPIs, quality data, measurement values, and reports
- Type and frequency of evaluations of data sets
- Number of interfaces and frequency of data exchange

These points affect the database in particular, whereby the *transaction volume* (effects on the CPU load and the interfaces of the system) and *data volume* (effects on the required memory capacity) aspects are taken into account. This means that both the processing power of the database system (usually in the form of additional processors) and the memory capacity (usually in the form of additional hard drives) should be suitable for flexible scaling.

Based on the suggested architecture in Sec. 7.1.2, the second bottleneck arises with growing quantity structures in the application server. Here, too, scaling can occur through adjustment of the processing power (i.e., additional processors), but the individual processes (applications) of the system can be allocated to several processing systems. The option to allocate the processes to several systems is a true advantage for system running and maintenance as well. With the example of the suggested architecture in Sec. 7.1.2, all components represented could be run on a common server in the simplest case. This architecture would lower the costs for a small system with small data quantities. At the other end of the scale, however, each of the software components represented could work in a separate server system. In order to be able to realize both extremes represented, the architecture must be appropriately flexible and as platform-independent as possible (see Sec. 7.1.3).

7.1.5 Flexible Adjustment versus Suitability for Updates

The only constant is change. Does this saying sound familiar? Correct—that was the introduction to the Sec. 7.1.4 of this chapter. But this statement is just as valid for this section. Changes to real production also require changes in the MES. These changes should be implemented as quickly as possible with low financial and organizational outlay. Despite this, the MES should be a standard product, that is, suitable for updates, stable, and ensured for the future. Thus it applies that the competing requirements of high flexibility and stability must be united in one system. This can be achieved only with extensive and complex *parameterization options* and simultaneous *suitability for updates*. Suitability for updates refers primarily to all core functions of the system, which must remain the same independently of the specific application. But what does *suitable for parameterization* mean? Is this just a number of switches, system parameters, or user profiles—or are other mechanisms also required? This question cannot be answered globally but must be answered specifically for individual modules and functions of the system:

- *Interfaces.* Various methods and technologies, such as OPC, telegram exchange via TCP/IP, remote function calls (RFC), message queues (e.g., MQSeries or Com+), Web services, and database interfaces based on views, have become popular for the technical development of software. Here, a flexible system should support various technologies, and the partners must agree on one of those technologies. The data exchanged are, however, hardly normed and are subject to frequent changes. Therefore, flexible tools such as programmable scripts are needed for configuration of the interface content. In order to guarantee suitability for updates for scripts as well, it must be ensured that already existing tools remain viable in the case of an update to the framework (e.g., scripting engine).
- *Main functions.* The main functions of the MES, such as resource management, fine planning, and machine data acquisition (MDA), should be available as *modules* of the overall system. This means that it is easy to carry out functional scaling. In the case of an update, individual modules also can be brought to a new status.
- *Partial functions.* Within these main functions are partial functions that can be activated or deactivated independently of

their application. Here, there is the option of configuration on the basis of *system parameters*. These parameters must be saved in an update-secure manner to avoid unpleasant surprises after software updates.

- *Project-specific data processing.* Data processing, such as the calculation of KPIs, is also strongly influenced by customer requirements. The use of scripts that are easy to change for the system supplier or that even can be adjusted by the customer is a tried and tested method. Here, too, the condition applies that scripts must be suitable for update. In the case of a software update, then, only the framework for running the scripts is changed to a new software status and not the scripts themselves.

- *User interfaces for standard functions.* It should at least be possible to adjust these interfaces in their look and feel to suit the needs of the user. For example, it should be possible to adjust a given corporate identity with predefined colors and a company logo globally. Saving the settings, such as selection of columns for a table view, and the default saving of the table should be user-specific.

- *Customer-specific user interfaces.* In some projects, not all needs of the customer can be covered with standardized interfaces. The software concept should allow for the project-specific creation of interfaces for exceptions. These interfaces created especially for a customer also must function after an update to the entire system.

- *Visualization via black diagrams/flowcharts.* Although standardized visualization with a generic approach (e.g., mapping out all machines/stations with the most important order-processing data) saves projection outlay, it is often too inflexible. A freely "parameterizable" visualization solution with the option to provide process visualization increases both the flexibility of the MES and its acceptance by users.

- *Reporting.* Meaningful and optically pleasant reports are the calling cards of the MES for company management, and this is why a highly flexible reporting system is needed. A set of standard reports should be present in the basic interface of the system. However, it must be possible for trained users to change these reports and for the reports to be used as templates for the creation of the user's own reports.

Software that is adaptable and also suitable for updates is absolutely realizable with modern IT tools. However, this comes at the cost of performance because of the necessary complexity. A simple system with few parameterization and setting options seems to be less expensive at first glance. The limitations can be seen only in the course of implementation or when the system is in use, when the first extensions are needed. Then the seemingly cheaper system also can develop into a money pit.

7.1.6 MES and Service-Oriented Architecture (SOA)

The basic idea of service-oriented architecture (SOA) aims to organize business processes into individual services. The client calls up a service for a defined task (i.e., order to the service), this order is then processed through the server, and the result (i.e., response from the server) is sent back to the client. The structure of the service (i.e., data structure for order and response) is managed in a common repository. A unique address (i.e., the server that provides the service) exists for every order to which an order can be sent.

The best-established technological SOA approach takes the form of Web services. The World Wide Web Consortium (W3C) has carried out an extensive standardization of Web services and data exchange using the Simple Object Access Protocol (SOAP, a protocol for data exchange via HTTP and TCP/IP) and therefore makes it possible to apply the technology in heterogeneous environments [W3C 2007] (Fig. 7.2).

The "Service Broker" depicted in the figure is required for company-wide or globally used services to inform an arbitrary client which server

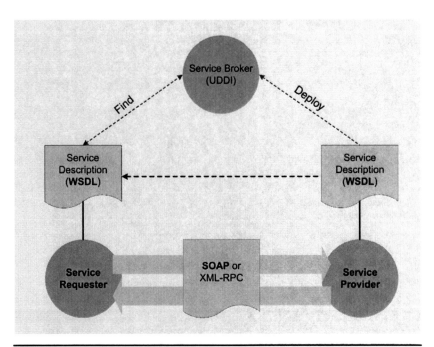

FIGURE 7.2 Concept of Web services with SOAP.

provides which services. These metadata, which describe a Web service, are exchanged with the aid of the Universal Description, Discovery, and Integration (UDDI) protocol. If these Web services are used only "locally," which is generally the case within an MES, or in connection with neighboring systems, the service broker can be omitted. The Web Service Description Language (WSDL) descriptions of the services are known to the clients in this case. The functions and the interface of a Web service are precisely defined in a WSDL file. With this information, the client can use the service provided. The data exchange itself is carried out via SOAP (see above; usually via HTTP, but other protocols are also possible) or via RPCs.

Using this architecture, it is possible to arrive at a situation where a *process* and the related data are mapped out only *once in the company's IT system,* and the software functions are still made available to all users in their specific context. Thus the desired integration of applications and production data (see Sec. 2.4.2) is achieved.

For example, internal interfaces such as the connection of MDA (machine data acquisition)/PDA (production data acquisition) terminals to a server can be implemented in a simple, flexible manner with Web services. Another example is the handover of attendance records for workers to the MES from a staff timekeeping record. Using these data, the MES can plan the resources that are actually present or carry out a plausibility test for order responses related to workers. By querying the data using an ID, this also can be rendered anonymous in the MES. In this example, the staff timekeeping system is the server, and the MES is the client for the Web services.

7.2 Database

7.2.1 Introduction

The database is the central and, from a technical point of view, also the most important element of the MES. Migration to another database system is very cumbersome and can hardly be carried out when the system is running. Thus it is important to decide on a specific database before the system is implemented. Each database requires a specific outlay for system maintenance when it is in use. For this, too, rules must be defined before the system is implemented, and resources with the necessary qualifications must be planned. Finally, the server should be coordinated for the expected resource load, and concepts for data security and archiving should arise in good time.

7.2.2 Resource Monitoring

In connection with the database server, four main resource reports can arise that must be evaluated and considered in the overall layout of the system:

Technical Aspects 131

- Pressure on the server's CPU caused by transactions
- Input-output (I/O) load caused by read and write access to the memory
- Booking in of data into the database via driver connections
- Memory requirements on the server's hard disk

The *CPU load* arises through booking processes, which are initiated externally to the database, and subsequent transactions; through queries and evaluations initiated by the user; and through internally triggered jobs for data maintenance or cyclic calculations. By assessing these influencing factors, it is possible to quantify the arising transaction volume and therefore the CPU load. In order to ensure the estimation, it is recommended to carry out load tests on a test platform, which should be as similar as possible to the target system. Scaling can occur either through the retrofitting of additional CPUs or through division into several instances (see Sec. 7.2.2). From the point of view of system architecture, the CPU load also can be reduced considerably by carrying out all data connection functions externally (e.g., on an application server) instead of within the database (in the form of stored procedures). Although this improves the scaling options, processing becomes slower and less secure. Unlike external processing of the procedure, the stored procedures function very effectively. They also require fewer data to be moved across interfaces.

The *I/O load* arises necessarily because of transactions and can hardly be influenced by software architecture. Here, hardware architecture must be adjusted as well as possible to the database system used.

Independent of the driver concept used (e.g., ODBC, ADO, or "native client"), one or several *transfer channels for the database* can be constructed from the interface adapters (see sample architecture in Sec. 7.1). However, the potential number of SQL statements used (e.g., INSERT or UPDATE) per time unit and channel is limited by serial processing. For example, an INSERT statement is only ended when the information is actually written to the hard drive of the server and the related response occurs via the driver. Only after processing has been confirmed via the connection is the next statement from the queue provided to the database for processing. In the case of a large number of measurement values, this causes the queue to lengthen and bookings to be delayed accordingly. This bottleneck therefore arises at the clients passing the data to the server and not at the server itself. However, for some databases, scaling through several simultaneously active connections is possible (e.g., for Oracle, but not universally).

The *memory requirement* of the application is established by the given requirements. In modern computer architectures, the hard disk capacity is not a serious problem with the introduction of a new system. However, if measurement values and production data are collected over months and years, a resource bottleneck also can arise here. Apart from absolute space requirements, an organizational problem also arises with data security in very large databases—a daily full security is no longer possible if the running time of the security exceeds 24 hours. In order to avoid these problems, intelligent archiving concepts are required to limit the size of the online database. An "alim" database keeps the system capable of performing and saves money.

7.2.3 Scaling the Database System

Scaling the processing power, that is, using several CPUs, is the simplest way of combating the bottlenecks described earlier. A condition for this is that the database supports a multiprocessor system in such a way as to ensure that the processors can be used as well as possible. A cost disadvantage arises not only through the hardware costs but also because of the licensing model of the database (processor license).

From the aforementioned problem of "booking in" data, three main requirements for design of the interface adaptor and the database arise:

- Ideally, the data are buffered at the source, for example, in the production's control system. However, these systems are usually not suitable for this. Therefore, the interface adapter definitely should develop the handover of data through SQL statements via a buffer (i.e., queue). Only this can result in data loss in cases of overloading or interruptions to connections be avoided.

- The data should be handed over including a time stamp (i.e., time of actual arising).

- The interface adapter and the driver concept used should allow for scaling of the connections, which are in use simultaneously.

- Only changed data should be transferred. Before transfer, the interface adapter must check every variable to see if there has been a change as compared with the value last transmitted. For particular values (e.g., based on official provisions), the booking of every value must be possible with a set time pattern (i.e., independently of a change). The optimization described therefore must be deactivatable selectively.

Alternatively, the data-collection concept can be changed so that the interface adapter does not transmit every value to the database but collects values, possibly compromised, and then undertakes a booking for a larger number of values (e.g., each minute). The disadvantage of this concept is that querying individual values is more difficult.

Technical Aspects

A possible approach to scaling of the database and increasing reliability is division into several instances, for example, into an online instance and an archive instance. If the situation requires it, these instances can be run on two computers. This almost doubles the resources available—*almost* because the comparison of both instances also causes a constant resource load. For very demanding systems, this concept can be extended with a third database instance (i.e., reporting instance).

7.2.4 Data Management and Archiving

Which data have to be saved for how long in the database? This question is hotly discussed in a project because only part of the data recorded must be maintained for a defined period of time for legal reasons. There are no such legal provisions for the other part, such as disruption reports from automatic systems without direct reference to the production process. Here, the problem generally arises that there is no responsible party for agreeing to the final deletion of the data. As a result, larger and larger data groups are created in the MES application over time, which can lead to disimprovements in performance and an increase in running costs (e.g., disk space and security media). Help can be provided by a multilevel archiving concept (Fig. 7.3), which can be adjusted to meet the needs of the project by means of flexible parameterization.

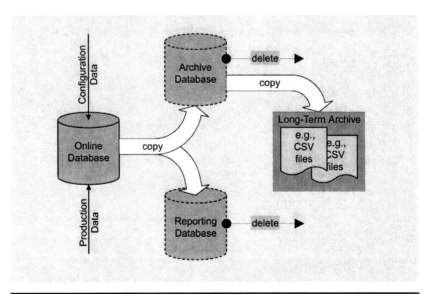

FIGURE 7.3 Multilevel data storage with several database instances and long-term archive.

7.2.5 Running Maintenance

A database generally requires regular maintenance (more than the other elements of a software system). Through the constant changing of the data set and data volume, the need for constant improvement arises. If you neglect the necessary maintenance measures, there is the threat of costly disruptions and data loss. Independent of the architecture selected and the database used, the following maintenance measures should be carried out regularly:

- Execution and checking of complete data security.
- Reorganization of the database and index maintenance with the integrated methods of the database system. Some databases only allow for maintenance measures in "office mode," so the application is not available for the duration of the maintenance measures. At required availability of 7 × 24 hours, it should be ensured that the necessary maintenance jobs can be carried out online.
- Monitoring of the data volume (i.e., memory requirements) and checking for unusual changes.
- Updating the database system with manufacturer-specific patches.
- Performance measurements and possibly optimization measures in the data model (e.g., introduction of a new index)

7.3 Interfaces with Other IT Systems

7.3.1 Overview

The interfaces of the MES with neighboring and subordinate systems, in accordance with the suggested architecture in Sec. 7.1.1, should be present as independent modules or software processes to achieve the required flexibility and scalability. Since the reference data to be exchanged, the integration of these data into the system, and the reports on the initiation of the data exchange (i.e., trigger) are very different, a *flexible interface* concept is needed. The use of a *script engine* that serves interfaces via a script that is easy to change is one such concept, for example. JavaScript has proven itself as a scripting language for such and similar applications because it is widespread and can be applied independently of the platform (Fig. 7.4).

7.3.2 Interface with Production

An increase in value creation is a primary goal of an MES. For application, the approach to date is sufficient to carry out planning that is not associated with the actual condition of the production and for

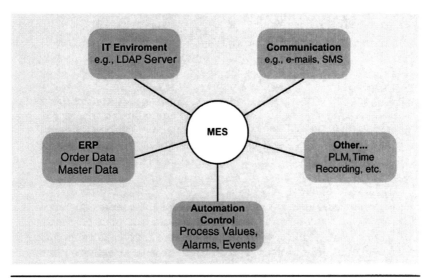

FIGURE 7.4 Interfaces of the MES with other IT systems in the overview.

nothing more. In fact, only a constant linking of the order and production data allows for an overall evaluation. Thus the data-related connection with production, that is, a vertical integration of the production data, becomes increasingly important.

Data that can be gained automatically by the control system are to be preferred to manually recorded data in any case. These data are unadulterated and, above all, available immediately and therefore improve the overall data quality of the MES. What use are the nicest evaluations and the best presentations of the results if the data on which everything is based are not reliable?

The data-related connection of all automatic or partly automatically controlled production or logistics areas (e.g., production, processing, materials handling and logistics controls, tool machines, robotic systems, etc.—hereafter referred to as *production controls*) is the basis for the successful implementation of an MES. However, something that sounds simple can mark the end before the beginning for some projects because the costs for networking and software adjustments are often too high. It becomes especially difficult when the production controls are only available to a very limited extent for necessary changes owing to high loads. The only way to keep this outlay for adjusting the infrastructure costs low is an across-the-board *standardization* of the production controls, especially for the interfaces to the MES. Ideally, a homogeneous system landscape from the point of view of the MES arises here, and new production controls can be connected to the MES with few work steps. The *plug and play* principle actually applies and not the interface chaos feared as per the motto "Plug and pray."

Connection via Process Visualization/ Control System Interface Adapter

At first glance, connection via process visualization, which is potentially already present on the machine or system, is a tempting solution. For applications with low data volumes and straightforward security and time-response requirements, this approach absolutely can be taken. Most modern visualization systems allow for the reproduction of data via an SQL connection, whereby the SQL statements can be embedded in the scripts of the application. Direct connection to the control system, in comparison, offers a better time response, higher data security, and lower maintenance outlay.

An additional interface adapter (see sample architecture in Sec. 7.1.2) that has been designed specially for the connection function offers some advantages. Larger data quantities can be transported in a shorter time, and maintenance occurs at a central point, which makes it easier to standardize the individual connections. With the aid of a central component for connection, a generic concept also can be realized, through which the outlay per couple partner is minimized. The connection of an additional interface, for example, can occur simply with the provision of name and Internet Protocol (IP) address. The entire data exchange then is developed in standardized form, whereby variables (e.g., for counted measurands or measured values) and events (e.g., got disruption reports) can be constructed in an automated manner within the MES.

Technologies

Today, mainly four variants for the connection of control systems to a management system exist on the market:

- *Openness, productivity, and collaboration (OPC)*. OPC was defined originally in order to solve the ever-returning problem of connecting PC-based applications, especially supervisory control and data acquisition (SCADA) and human-machine interface (HMI) systems to the process periphery as a unit. In May 1995, the newly founded OPC Task Force set up for this purpose, met for the first time. In December of the same year, the first draft specification of Data Access 1.0 was published. Today, a decade later, OPC has established a standard that is valid worldwide for data and information exchange of software components. With over 7500 OPC products and millions of installations in a wide range of industry branches, the OPC initiative can be judged a complete success. For a long time, OPC has not just been applied instead of proprietary communication drivers for the connection of SCADA systems and visualization programs to the process periphery. Process management systems, PC-based controls, MES, and even ERP systems can no longer be imagined today without OPC interfaces.

Nowadays, it is not just process data or individual parameters that are transmitted via the OPC interface; entire material management documents, parameter sets, control sequences, video signals, and driver programs are transported via OPC.

The roots of OPC are closely linked with Microsoft's Windows operating system. The original meaning of *OLE for process control* comes from the Microsoft object linking and embedding (OLE) technology of the 1990s. Soon OLE was replaced by component object model (COM) and distributed COM. At least since the extension with extensible markup language (XML) and Web services within OPC data exchange and OPC XML DA specification, the original meaning of OPC has no longer been correct. Thus, today, OPC stands for "openness, productivity, and collaboration" and reflects to a lesser extent the link with a particular technology and to a greater extent the characteristics of an open, interoperable, and productive OPC interface.

With the distributed component object model (DCOM) as a technological base, the use of OPC until now has been limited to the automation level and to Microsoft Windows platforms within a company's network. Web services and XML remove this limit, whereby data and information remain isolated behind a company firewall and cannot be published for communication that reaches across platforms and for the Internet. The vision of global OPC interoperability and the migration to a uniform public Internet architecture/platform for the information flow from the factory hall to company management thus has become a reality.

The continually growing number of members of the OPC Foundation and the ongoing extension and modernization of the specifications show that OPC has established itself as a true standard for communication to the control level. Therefore, a suitable OPC server for the connection of any production controls generally can be found on the market. In the course of the past few years, the following standards have been specified by the OPC Foundation [OPC 2008]:

- *OPC DA (data access)*. Specification for the transmission of real-time values. OPC DA was the first OPC specification and can be found in many current products, from programmable logic controller (PLC) systems (control that is programmable from memory) to process visualizations. Most products comply with the specification OPC DA 2.0; the latest version of the specification is 3.0.

- *OPC A/E (alarms and events)*. Specification for the transmission of alarms and events.

- *OPC HDA (historical data access)*. Specification for the transmission of historical values.

- *OPC DX (data exchange)*. Specification for direct communication between OPC servers.
- *OPC command*. Specification for the execution of orders.
- *OPC XML DA*. Specification for XML-based transfer of real-time values. The disadvantage of this and the aforementioned specifications was the DCOM technology, which was linked to Microsoft platforms. Therefore, a platform-independent variant, namely, Unified Architecture, was announced shortly after these specifications were drawn up. The distribution of OPC XML DA is at an accordingly low level.
- *OPC UA (unified architecture)*. Specification that unites all specifications to date in a platform-independent manner (without DCOM technology). The core of this specification describes a service-oriented architecture with Web services (SOA; see Sec. 7.1.5) and thereby follows the current IT trend. For data exchange, a binary variant (OPC UA binary) is available in addition to the Web service variant (via HTTP/SOAP) with which a considerably better performance can be achieved because of the low overhead. Furthermore, OPC UA binary uses the fewest resources because an XML parser, SOAP, and HTTP are not necessary.
- *Modbus TCP*. The Modbus protocol is a communication protocol that is based on a master/slave and client/server architecture. Alongside solutions such as OPC and Profinet, it has emerged as an established standard for communication via Ethernet-TCP/IP in automation technology. The basis for this is a stable specification, available basic technology, and a large number of industrial serial models such as controls in any performances class.

  Unlike distributed automation solutions such as Profinet, fieldbus-on Ethernet solutions such as Mobus/TCP are characterized by the fact that the current fieldbus protocol is largely retained in an unchanged format, and Ethernet-TCP/IP has been adopted as the new transmission technology. Significant advantages of these systems lie in the fact that the specifications have been stable for years, and implementation does not require any fundamental rethinking for users. "Less is more" is the motto for these solutions, which are not designed for distributed automation but rather for a fast, reliable transfer of data between automation devices and field devices.

  Accordingly, Ethernet-TCP/IP was adopted as a further transmission technology for the Modbus RTU protocol, which has been known since 1979. The Modbus services that have been tried and tested since the original variant, such as the reading and writing of address spaces, have been retained in an unchanged form and are mapped out for TCP/IP as transmission media. The disadvantage of RTU was that only a 1:1 ratio was possible between the communicating stations because of the medium (serial interface). With TCP, each party has a unique address. Through this extension, a 1:n relationship with the Ethernet medium is possible [MODBUS 2007].

- *Profinet IO.* Profinet IO is the latest specification of Profibus, and it builds on the tried and tested Profibus DP function model. Here, it uses Fast Ethernet technology as a physical transfer medium. The system is tailored for the fast transfer of I/O data and at the same time offers a transfer option for requirement data and parameters as well as IT functions; existing know-how about Profibus DP can continue to be used. As with Profibus DP, the decentral field devices with Profinet IO are linked to the projection tool by means of a device description. The characteristics of the field device (Profinet IO device) are described by the device manufacturer in a general station description (GSD) file. The periphery signals are read cyclically in the SPS, processed there, and then output once more to the field devices. With Profinet IO, a provider/consumer model is used in contrast to the master/slave process used with Profibus; this provider/consumer model supports communication relationships between users with the same rights within Ethernet. A significant characteristic here is that the provider transmits its data without a request from the communication partner. As well as cyclic reference data, Profinet offers additional functions for the transmission of diagnoses, parameterizations, and alarms. As is the case with Profibus DP, the devices are classified according to their typical tasks in Profinet IO [PNO 2007].

- *Ethernet TCP/IP.* The telegram-oriented exchange of data via the Transmission Control Protocol/Internet Protocol (TCP/IP) is the least standardized but also the most flexible method of data exchange. Communication components for a TCP/IP connection via Ethernet are available for almost all SPS systems. For PC-based systems, the conditions are also "on board." The aforementioned technologies do not just regulate the transport of the data, but they also define rules for the exchange of particular content and provide mechanisms for this. In the case of a simple TCP/IP communication, the user must carry out this standardization. The best approach is specification of the most needed telegrams in an *interface description*. Here, the structures and contents of the required telegrams (e.g., "process value" telegram, "alarm" telegram, or "order data" telegram) and the data-exchange rules (e.g., fixed or variable telegram length, acknowledgment behavior, establishment of connections and startup behavior, etc.) are established in a binding manner. Both sides (i.e., production control and the MES) thereby receive clear provisions with regard to communication. The advantages of this solution are flexibility and the option to be able to exchange all data in a report-oriented manner—a polling process such as is usual with the application of OPC thus can be avoided. On the other hand, the high (one-time) outlay for creation of the interface description and the outlay that must be driven on the production-control side to implement this standardized data exchange are disadvantages. Especially if machines and systems, including the control system, are bought ready made, such a standard, which is seen by the machine manufacturer as a customer-specific standard, usually can be implemented only with difficulty. Therefore, this solution is suitable only for larger projects with a large number of control systems. The one-time outlay can be allocated to many systems and is more than balanced by the standardizing effect that is created.

7.3.3 Interface with an ERP System

Overview

If the MES is not implemented as a standalone production management system but is implemented in combination with an ERP system, an interface also must exist between both these systems. The type and extent of the data exchanged strongly depends on the systems used. In particular, the question, "What data quantities will be managed in which system?" must be answered before the interface can be defined. In Fig. 7.5, the content of the interface is mapped out for typical ERP system/MES task sharing.

For all *transaction data* (data that arise in order management or productions flow), the requirement that data traffic should occur asynchronously applies. *Asynchronous* means that the process that transfers the data does not have to wait on the recipient to be able to execute its own program sequence. Thus it is guaranteed that the systems continue to operate independently of each other in the case of disruptions. This gives rise to a further requirement that also improves the reliability of the interface: the data for the transfer must be buffered in the transmitting system to avoid data loss in the case of a disruption.

Master data that are to be aligned between the systems are either transmitted via intelligent import functions or are created

Technical Aspects 141

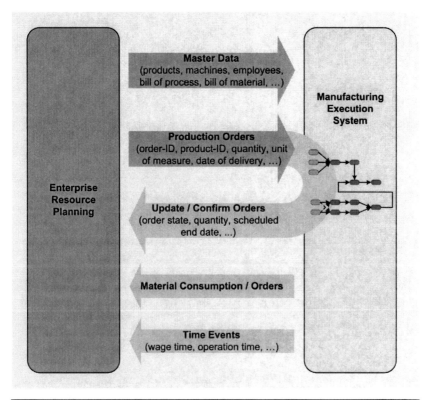

FIGURE 7.5 Content of the interface between an MES and an ERP system.

dynamically in the target system. For example, if an article that is not yet known in the MES is transmitted with an order, a data set for the master data of this article can be created dynamically in the MES. The alignment of the entire data set can, if required, occur at a later point in time.

Technologies
Many ERP systems offer several technical variants for the exchange of data. The most important of these are described in brief below. MES also should support these "market typical" technologies and techniques to avoid expensive, high-risk additional developments.

- *Remote procedure call (RPC)*. Communication is created through calling up a "remote" (i.e., located on another computer) procedure; see below for an explanation of the technique. A special form of RPC was developed by SAP with the remote function call (RFC). Under this concept, SAP has summarized its own protocols and interfaces for the execution of the function call. We distinguish among *synchronous, asynchronous,* and *queued* RFCs (to guarantee a defined processing sequence of execution across an order queue). For applications that wish to link with an SAP system, libraries are provided for various programming languages/run-time environments.

- *Web services.* Using a Web service, an MES (in the role of the client) can transmit, for example, responses to a production order to the ERP system (as a server). See Sec. 7.1.6 for technical description of Web services.

- *Database interface.* The MES retrieves the required data by means of a database interface through direct access to the ERP system database, or vice versa. This "primitive solution" demonstrates considerable disadvantages: the simple form of the data exchange does not allow for any true standardization. Security with regard to unauthorized access and transaction security for the data exchange are hardly guaranteed. Furthermore, the time response of this solution is not perfect because the data exchange cannot be run via events but takes place within a polling process. In order to be able to offset these disadvantages even partially, the following precautions absolutely must be taken:

  - Access must not be to the tables of the other system but must occur via *views.* These are specially created for defined access. The data are provided at the time of the query, so also with the access. This means that the actual data are protected from undesired changes. In addition, no integration into the complex data model of the communication partner must follow because data from several tables can be summarized in a user-friendly manner in a view.

  - A specific database user profile should be set up for access that receives only the minimum rights needed for data exchange (e.g., read access to defined views).

- *File interface.* A file is written by a commonly used directory (share) and read by another system. Generally, the file is structured in its makeup, for example, in the form of a comma/character-separated values (CSV) or XML file. After the recipient has read the file, this is deleted or renamed. This means that the way is clear for the next transmission. One example of this also very simple method of communication is the transmission of production orders from ERP system to MES. The disadvantages of the solution generally are identical to the database interface problems described earlier.

Technical Aspects 143

> **Remote Procedure Call (RPC)**
> Data are exchanged by calling up a procedure that is located on a remote system. This is a communication technique via a network that is being used. RPC was developed originally by Sun Microsystems for the Network File System (NFS). The idea is based on the client/server model and should make it possible to use program functions collectively across computer borders. We differentiate between synchronous and asynchronous RPC. In the case of synchronous RPC, the client making the call waits with the execution of additional programs until a response for the procedure has been received from the server. With asynchronous RPC, on the other hand, the client does not wait for the response and can continue to run the program code. Another RPC variant is XML RPC. The data to be transmitted are recorded in an XML document and transmitted via an HTTP connection (see also Sec. 7.1.6).

7.3.4 Interface with the IT Infrastructure

For maintenance of the entire IT infrastructure of a company, it is important to use technology and components that are as similar as possible in all systems. One example is the central administrating of system users and their passwords. For this purpose, every user of software systems is recorded in a central directory with his or her user ID (e.g., encrypted staff number) and the corresponding password. This directory, for example, is represented by a Lightweight Directory Access Protocol (LDAP) server. In order to avoid multiple maintenance of users and passwords (which is a real problem in large companies), an MES must be capable of verifying a user login via the LDAP server. The LDAP server receives the login information from the MES and reports back whether the login of the user is valid. The system-specific rights of the user are recorded in the MES. An additional example of a directory service is the Microsoft Active Directory (directory service on a Windows server).

Other interfaces with the general company IT are

- Alignment of master data with a master data management (MDM) system.
- Synchronization of the system time with a time server.
- Download of updates/patches for the operating system from an update server.
- Preparation (and possible forwarding) of system reports in a log file for central evaluation in an IT control center.

- Forwarding (or also query from the company's own system) of system reports to an IT control center by means of the Simple Network Management Protocol (SNMP).
- Central filing of reports on a share with uploading and downloading options.

7.3.5 Interface with Communication Systems

Modern communication technologies such as e-mail and Short Message Service (SMS) offer advantages for the functions of an MES with regard to information distribution. The information can be sent to recipients quickly and precisely. Therefore, the MES should, for example, support interfaces to SMS servers, wireless paging systems, paging systems, telephone systems (voice output), and/or e-mail servers.

For a technical description of these interfaces, the statements in Sec. 7.3.3 are valid.

7.3.6 Other Interfaces

In the context of company management level and production management level, there are still ranges of further interfaces that affect MES. Applications that require data from the MES or supply the MES are, for example, product lifecycle management (PLM) systems and time-measurement systems.

For a technical description of these interfaces, the statements in Sec. 7.3.3 are valid.

7.4 User Interfaces

7.4.1 Usage and Visualization

Requirements

From the point of view of the user, user interfaces and reports are the most important and often the only points of contact with an MES. It depends considerably on the entire usage of the system how comfortable users feel with the system and how well they can use the system in their task areas. The usage—and thus the cost-effectiveness of the MES—therefore is closely linked with acceptance.

Assuming that the product implemented is a standard system and therefore that there have been no customer-specific developments, the user interface is also standardized and is not adjusted to suit the special requirements of the customer. However, this is a clear contradiction of the requirements "feel comfortable" and "optimal usage in a specific environment." The following characteristics of the user interface are suitable for combating this problem:

- The standard user interfaces (user interfaces that are linked permanently with the system, e.g., in the form of tables or input interfaces) should be suitable for global adjustment via a *style concept* to the requirements of the customer. For example, a customer logo, background colors, fonts, and table layouts are saved in one style. The general appearance of the application thus can be adjusted to the requirements of the customer.
- It should be possible to adjust the content and functions of the standard user interfaces as group- and user-specifically as possible. This configuration includes the free choice and sequence of fields in tables, the standard sorting of tables, the displaying or fading out of main menus and submenus, and the size of windows and frames. The settings, once made, must be saved in connection with the user account in order to present the interface to which the user is accustomed on his or her next login.
- As well as the standard user interfaces mentioned above, the MES also should have access to a fully freely programmable visualization, comparable with the functions of process visualization. This means that a fully graphical user interface with systems and process images can be realized that, for example, is of particular advantage for repair and maintenance interests and a control center. For the creation of these graphics-based views, a convenient tool must be available. An object-oriented structure with the option for managing a project-specific library can speed up creation of the views considerably. The maintenance (and if necessary, the new creation) of these views should be the responsibility of trained users on the customer side. This means that the customer is able to optimize the system on a regular basis and adjust it to suit the continual changes in the company's production.

Another requirement is that *information from the system is available as constantly and currently as possible and in any place* (often also outside of and in different locations from the company). This requirement can only be fulfilled sensibly with a Web application, that is, through use of the MES by means of a standard Internet browser. The advantages and disadvantages of a Web application are examined more closely in the section on "Technologies".

Visualization via Large Displays and Andon Boards

The use of large displays and andon boards (see concept explanation in the box below) also ranks among the tasks of the MES and is part of the user interface. Through the recent trend in the direction of lean production, such high-ranking visualization and motivation systems are becoming more and more important.

The most widespread are large displays based on multicolored light-emitting diodes (LEDs). This solution also best fulfils the size (often more than a meter in diagonal) and brightness requirements. In opposition are the high acquisition costs, low resolution, and high level of energy consumption, which also leads to high maintenance costs.

With the slide in prices for liquid-crystal displays (LCDs), these systems are being used more and more frequently as large displays. The advantages are high resolution (image masks must not be designed specially for the medium), low acquisition costs, and the option to run the models as Web clients.

Andon Board

The concept *andon board* originated in Japan and is a system for the triggering of improvement measures. With the aid of andon systems, workers can trigger optical and/or acoustic signals in the case of quality problems or disruptions, for example.

In today's language usage, an andon board is a display system that is generally located under the ceiling of the hall and displays status information from the production area in a manner that is highly visible. The workers from the area are informed about the current production quantities (i.e., target and actual values for the current shift and predicted quantities for the end of the shift), about serious disruptions in the production facilities, or about quality problems via the andon board. The circumstances depicted are either transmitted directly from production control (e.g., quantity measurement or run rates) or are triggered directly by the workers themselves (e.g., reporting a quality problem via a release-line system).

Mobile Solutions

Mobile end devices are well established because of their advantages, such as data processing independently of location and permanent wireless accessibility in today's usual business day. In industry, too, mobile computing is becoming more and more widely accepted. A mobile end device such as a personal digital assistant (PDA) or "smartphone" (i.e., a combination mobile phone and PDA) also offers some advantages as a client system for an MES compared with permanently installed clients. Some ideas on this topic are as follows:

- A machine or system user who needs to use several machines/ systems can bring his or her interface directly to the location of the events. The practice until now of installing a terminal at every machine (cost-intensive) or covering large distances thus is a thing of the past.

- The manager can gain an overview of the production status regardless of his or her location at any time. With a smartphone, the manager also can gather information immediately where needed or implement measures.

- A maintenance and repair worker who has, for example, been instructed to resolve a disruption to a system can view the current system status before he or she has reached the system. In addition, he or she can recognize and assess effects (consequent disruptions) on neighboring areas more easily.

However, the devices available today do not offer the resolution of a PC screen, so a 1:1 display of the content is possible only at a cost to the ergonomics. Parts of the user interface that are intended for use via mobile end devices also should be designed specially for this application; that is, a part of the user interface must be developed twice.

Technologies

A fundamental decision must be made between the "rich client" (also known as the "fat client")—client software on the MES is installed on every client PC—and "thin client"—without special client software—variants. For the thin client, there are two competing concepts:

- Use of a terminal server where a remote MES client is provided via special software (e.g., Citrix, Remote Desktop, or X-Server)

- Mapping out of the user interface in the form of a Web solution with controls via a standard Internet browser

The great advantage of a true thin client, which is that no special software needs to be installed on the PC, is still in opposition to a lower level of user comfort for most Web solutions. If you intend to use both advantages, you also should implement both variants in one system:

- For all monitoring and information functions, a thin client is the optimal solution. The required information is available at every PC workplace that is connected to the company network. Queries and reports also can be carried out from home via secure access. The use of mobile end devices (see "Mobile Solutions") is also only possible in a sensible manner with a Web client concept.

- Complex user functions that also require intelligence at the client end (e.g., a Gantt plan for the production control center) can be implemented as rich clients. Since in this case there is a manageable number of permanently allocated workplaces, the argument for the necessary installation and maintenance outlay does not make as much of an impact.

For realization of a real Web solution, various basic technologies are available. A relatively new concept is known as *asynchronous*

JavaScript and XML (AJAX), and it combines the advantages of a slim Web solution (i.e., no applets or other objects that need to be loaded) with the agility of a rich client. To put it simply, only changed data are exchanged between the client and the server, and only changed elements of the views used are updated. Libraries based on this technology are already to be found on the market, and they contain intelligent user objects (e.g., list box with a multiselect function). The development of application can be expedited considerably through the use of such libraries. With the aid of AJAX, it is also possible to realize graphic visualization. However, independent graphics formats that are as text-based as possible and have a convenient tool for creating graphics are better suited to this purpose.

These requirements are met, for example, by scalable vector graphics (SVG), and this approach also offers the option to scale to the run time. This means that a zoom function can be realized in the graphic views. An SVG file is mapped out in XML, which allows for editing with simple text editors and also for generic architectures. Flash is the oldest and still the most up-to-date technology for interactive and animated Web sites. Formerly developed by Macromedia and then taken over by Adobe, the format, which is based on vector images, offers primarily very good support for multimedia content and is therefore especially suitable for animations and advertising messages. The file name extension for Flash files is SWF, which stands for "small web format" or "shock wave flash." The most common browsers support both SVG and Flash by means of plug-ins or through run-time environments that are already integrated.

Another technology in this context and also an alternative to the aforementioned SVG and Flash formats comes in the form of XAML from Microsoft. XAML (eXtensible Application Markup Language) is a language that is also mapped out in XML for the description and creation of surfaces for the Windows Presentation Foundation (WPF) and is also a core component of the .NET-3.0-API of Windows Vista. XAML actually was designed for the Windows platforms (XP and Vista), but it should be possible to use this system on other operating systems (e.g., Linux) using Silverlight. Silverlight is a technology, also developed by Microsoft, which was presented for the first time at a conference in April 2007. It is a portable run-time environment for applications defined in XAML in all relevant Internet browsers (i.e., Firefox, Opera, Safari, and Microsoft Internet Explorer). In the first stage, applications created in AJAX or JavaScript are supported. In Version 2.0, connection to the .NET framework is intended.

7.4.2 Reporting

An MES must be capable of providing reliable reports for different user groups. These include shift reports per production area to provide production management with information, disruption reports for maintenance and repair, quality reports for quality assurance, and

KPI reports for management. These reports usually are laid out in the medium and long term and should support workers in their daily work. Only a limitation of the data to be evaluated must arise by means of a flexible filter (at least production area, time domain, and article). The basic structure of the report is retained.

The second requirement for reporting is a spontaneous nature; for example, after a change of product systems, for the introduction of new products, or for the monitoring of burning issues, new reports are needed in the very short term whose contents are application-specific and therefore can hardly be standardized. These ad hoc reports therefore must be created quickly on the user's request and lose their reason for existing just a few days or weeks later. The fundamental evaluations also only affect short periods.

The creation of standard reports that should cover most normal requirements and ad hoc reports are two tasks that represent different requirements for the reporting system. Therefore, in the first step, it should be assessed whether the two types are really needed and whether both tasks must be covered by the MES. Often reporting or business intelligence systems (see concept explanation in the box below) already exist within the company that can then take over the ad hoc reporting for the MES. The advantage of this variant is that user know-how for the creation of the reports already exists within the company. This helps to save training and running costs. However, a condition for the use of a reporting system that is external from the point of view of the MES is that the MES database generally is available to external systems and especially that the data structures are documented to a sufficient extent. The data also must be transmitted to the repository/data warehouse of the business intelligence system in cycles that are as short as possible so as to ensure than an ad hoc report actually contains the recent past (ideally, until now).

Independent of the question of whether an integrated or external reporting system is to be used, it must be possible for the user to maintain the reports with ease. This means that the reporting system must come with a convenient designer for creation and maintenance of the reports.

Business Intelligence

This concept includes the analysis of company data (business) and the knowledge gained from it (intelligence). Three main steps are included: collection of the raw data, creation of connections, and drawing conclusions. The first step, the data collection, is carried out by the MES. For the creation of reliable connections, a business intelligence system has its own repository or data warehouse that contains multidimensional analyses based on statistical methods such as online analytical processing (OLAP) or data mining (evaluation of a data set with the discovery of distinctive features by means of pattern recognition). Transformation

> of the raw data collected into a format that can be evaluated is the core task of business intelligence. Reporting is a partial aspect of this third step, namely, the drawing of conclusions. For this purpose, methods of information management are also used.

The reports generally are created through manual triggering by a user. However, especially for standard reports, it makes sense to have the option to use automatic, timed report creation. A case where this situation arises often is the shift report that is intended to provide the responsible parties with a quick overview of production status. This report can be initiated through the shift-end event and then is available immediately for the daily meeting. Distribution of the report can occur through saving it in a defined directory in the simplest case. A distribution option via e-mail can increase user comfort and improve usage (see Sec. 7.4.3) (Fig. 7.6).

7.4.3 Automated Information Distribution

Modern communication technologies, especially e-mail and SMS, not only change our private communication behavior but are also coming to dominate information exchange within companies to a larger and larger extent. Unlike conventional methods of gathering information from a

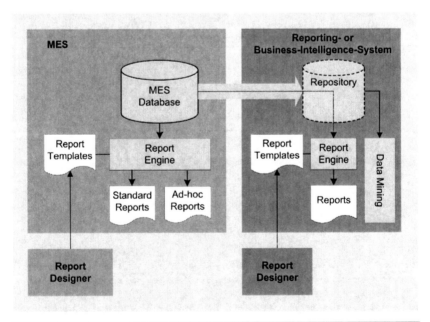

FIGURE 7.6 MES with integrated and/or external reporting or a business intelligence system.

display system, these new communication technologies have the advantage that the user is automatically informed about important events through the system. This means that not only does the user inform himself or herself in the MES system, but the MES also informs the user proactively. The proactive distribution of important information through the MES is especially useful in connection with mobile end devices (see also Sec. 7.4.1). From the numerous options for transmitting information, the three most common are examined in further detail below:

- *E-mail.* It should be possible to configure the content of e-mail using templates. Placeholders in these templates are replaced with a real value (e.g., OEE of the last shift) when the e-mail is transmitted. In order to avoid unleashing an undesired flood of e-mails, it must be possible for the recipient to edit the trigger to send the e-mail:
  - No sending in the case of absence (e.g., holidays)
  - Sending depending on a definable limit (e.g., only when OEE < 95%)
  - Sending with selectable time and rhythm (e.g., Monday–Friday 14:30)
    In addition to this information received directly in the text of the message, the e-mail also offers the option to send reports as attachments. Here, these must be created automatically and at a certain time by the MES. This means that the information content can be increased further.
- *SMS.* For sending information via SMS, the statements in the e-mail section are valid in turn. SMS is better suited to short messages (and in addition, no attachment can be sent) that are sent in an event-controlled manner. The typical application case is the informing of maintenance after a disruption has arisen. It should be noted that no defined transmission times could be guaranteed for SMS, which has a limiting effect. Therefore, this method is not suitable for critical processes.
- *Voice message via telephone.* The applications lie in the event-driven transmission of information, similarly to the MES. A special speech synthesis software model is needed to transform the text into a voice message. Independent of this module, the texts should contain no accented characters or abbreviations, where possible, to guarantee comprehensibility.

7.5 Summary

At the beginning, this chapter described two general approaches for the architecture of an MES whereby a suggestion, namely, the database-centered approach, is subsequently examined in more detail. Here, the central components of the system were explained. Essential characteristics of a modern MES, such as platform independence and scalability, were explained. The basis of innovative communication mechanisms, such as the OPC UA, is service-oriented architecture (SOA). This approach is also valid for the architecture and communication mechanisms of an MES.

The core of an MES is formed by a long-term database. The conditions for this core and the technologies necessary were described. Here, measures for archiving play as important a role as continual maintenance of corresponding systems; only through this can problem-free use of the MES be guaranteed.

Finally came a detailed view of interfaces, both with other systems and with the user. Various technologies and communication mechanisms that are used within the scope of an MES were examined.

CHAPTER 8
Evaluation of the Cost-Effectiveness of MES

8.1 General Information on Cost-Effectiveness

8.1.1 Calculation of Cost-Effectiveness

The concept of *cost-effectiveness* generally denotes measures for efficiency and/or rational operation with scarce resources. It is defined as the ratio between a result achieved (return = usage) and the use of resources required for this (outlay = costs).

Cost-effectiveness can be increased by striving for the most favorable ratio between target achievement and resource usage. In practice, the cost-effectiveness of a company, implementation of an information technology (IT) system, or product development can be established in three ways:

- Target–actual value comparison
- Comparison with other companies or measures
- Comparison between different points in time

In the simplest case, cost-effectiveness can be measured from the ratio between quantifiable costs and resulting profits. Thus a measure is cost-effective if the sales within a certain observation period are higher than the costs. Exact statements on this topic generally can be made only in retrospect. However, sometimes prognosis in advance of an investment is of much greater importance than controlling in hindsight. Still, the mathematically precise results of controlling are essential for sustainable prognoses.

The decision for or against a new investment [e.g., implementation of a manufacturing execution system (MES)] is made based mainly on improvements in cost-effectiveness that can be expected. In these cases, both the necessary investments (e.g., licensing fees, installation costs, training outlay for employees, etc.) and the predicted improvements (e.g., delivery reliability, processing time, machine utilization, etc.) should be measured reliably beforehand.

8.1.2 Comparative Cost Method

The *comparative cost method* is a static investment calculation process that allows several alternative investments to be compared. Here, the overall costs of the alternatives are determined, and the most economical is selected.

The overall costs are calculated based on the fixed and variable costs. Cost comparison calculation considers the average costs for a period. These capital costs arise from the calculated depreciation and interest.

This process therefore is very suitable for comparing several existing alternatives. However, it does not allow a fundamental decision for or against an investment as such to be reached. Furthermore, both the variable and the running costs for the implementation of an IT system are difficult to predict.

8.1.3 Value-Benefit Analysis

Similarly to the cost comparison calculation, *value-benefit analysis* can help to select the suitable option from a range of several alternatives that are difficult to compare with each other. Here, the alternatives must be parameterized and mapped on consequences that also can be parameterized. The analysis assumes that the decision maker prefers alternatives that will bring the greatest benefit. Unlike the cost comparison calculation discussed earlier, this is a suitable approach if there are "soft" criteria (i.e., criteria that cannot be represented as monetary values or figures) on the basis of which a decision must be made in favor of one of several alternatives.

The process of a value-benefit analysis can be represented as follows:

1. Clear definition of the investment decision goals and establishment of the most important goals to be achieved
2. Identification and listing of the goals as selection criteria
3. Weighting of the goals (total = 100 percent)
4. Assessment of goal achievement for every alternative investment object using an established scale
5. Calculation of weighted goal achievement per investment alternative
6. Summation of weighted goal achievement per investment alternative
7. Selection of the best investment alternative and creation of a ranking for the alternatives based on the weighted points achieved

It is apparent that the evaluation criteria and their weightings are of a subjective nature. Therefore, although value-benefit analysis is definitely a powerful instrument for every company when it comes to decision making, mathematically founded statements are not possible with this approach. This type of analysis makes sense particularly at the start of an evaluation process when financial aspects still play a subordinate role. However, it is ill-suited for establishing the cost-effectiveness of an investment in the sense of Sec. 8.1.1.

8.1.4 Performance Measurement

Performance measurement is understood as the construction and application of indicators of usually several dimensions that are used for assessing the effectiveness and efficiency of the performance and performance potential of various objects in a company, the so-called performance levels (e.g., organizational units of varying size, employees, and processes). In addition, more object-related and across-the-board communication processes and increased employee motivation can be brought about, and additional learning effects can be created using this method.

Performance measurement is a process involving the qualification and evaluation of the target achievement of organizational units, employees, and/or processes. Thus it involves the process of measuring and assessing performance, whereby not only the simple measurement but also the analysis of the performance results are meant.

Performance measurement is not limited to quantitative indicators alone. Quite the opposite: Qualitative indicators gain importance and are at least equal to quantitative values. Here, the more strategic the object assessed, the more qualitative indicators are used and consulted for decision making. There are several instruments within performance measurement. One of the most often used of these is the *balanced scorecard*. Because of its composition, it is always focused on the means-ends rationality, that is, efficiency [KAPLAN NORTON 1992].

8.1.5 Total Cost of Ownership

A possible instrument for cost monitoring for information systems that reaches across phases and includes every aspect is the *total cost of ownership* (TCO) concept developed in the 1980s by the Gartner Group. This approach is used to help consumers and companies estimate all costs arising from investment goods, especially within IT (e.g., software and hardware). The idea is to reach a calculation that includes not only the procurement costs but also all aspects of later components. Thus known cost drivers or hidden costs can be identified in advance of an investment. The most important basis for a deeper understanding of the TCO is the differentiation between direct and indirect costs.

Since its introduction, the TCO approach has become increasingly important for companies in their efforts to assess IT investments.

On the one hand, the company's own TCO is of interest with regard to the investment in an MES (TCO on the cost side) that is being considered. On the other, MES systems influence the TCO positively for machines, for example, at the user side on the basis of specific MES functions (TCO on the user side).

8.2 General Information on Evaluation

8.2.1 Assessing Cost-Effectiveness in Practice

The scientific cost-effectiveness assessment of an MES is nonexistent de facto. Often suppliers of corresponding systems explain this lack by saying that a formal measurement is not necessary because the systems have existed for so long and have established themselves on the market, which quasi-equals an implicit assessment. Such an argument is certainly open to criticism. On the user side, especially in smaller companies, the reason is mostly that the time and personnel resources are lacking to carry out a complex assessment. However, it is also often the case that the necessary understanding for a monetary assessment and its complexity is lacking, and therefore, the decision is made to do without an assessment.

Large companies often understand investment in information systems as strategic investment. Therefore, carrying out a cost-effectiveness assessment is viewed in advance as unnecessary. Also, in the past, when enterprise resource planning (ERP) systems were introduced, this aspect was dealt with rather freely because these systems were considered to be an urgent necessity for the strategic decision making level. When computer-aided design (CAD) was introduced and the change to intelligent automation systems was made, readiness to invest was just as rapid at the decision making level because the rationalization aspect for CAD was obvious, and the introduction of automation systems was inevitable.

It was more difficult to win the decision making level of a company over to quality assurance systems. It was mostly external pressure through applicable norms and directives [ISO 9000] that made such an investment possible.

It is much more difficult to convince business management of the necessity to introduce an integrated MES. There are several reasons for this: First, it is asserted over and over again that the IT requirements of production can be covered with an ERP system and that existing isolated solutions are sufficient. Second, the company initially sees only the investment necessary for such systems. The resulting positive benefits that lead to an improvement in the economic sense are not apparent at first glance.

Fundamentally, every individual case must be looked at more closely. It definitely may be that an ERP system is sufficient for production companies with low production depth and without in-house production of subarticles and pure assembly. However, as soon as there are complex, expensive products with a large production depth, with a large number of subarticles and variants produced in-house, and with a wide range of orders to be processed "simultaneously," an ERP system is no longer sufficient. The event/real-time-related planning, execution, and monitoring of orders becomes more important. These tasks can be fulfilled only by an integrated MES.

When deciding on an integrated MES, however, it is also necessary to consider that most production companies already have isolated solutions for individual function divisions. If these solutions are to cover the needs satisfactorily in any form, the cost-benefit aspect of such a system must be studied very precisely. Then it may be that a solution integrating further components in the "best of breed" sense may be the best solution.

8.2.2 Rationalization Measures in Production

There are several strategic initiatives and approaches aimed at recognizing, avoiding, eliminating, or reducing sources of loss in production. These sources of loss largely concern time consumption in the value-creation process. The existing approaches are explained in brief below.

Toyota Production System

The Toyota production system (TPS) is a production process for mass production developed by Toyota in the past few decades and subject to constant improvement. It links the productivity of mass production with the quality of plant manufacturing. The goal is production with the lowest possible waste of resources of any kind. Information regarding what should be produced in what quantity is forwarded to the upstream point from the downstream point using *kanban* cards. Thus only the quantities that are actually needed are produced. The process is also known as *pull production (pull principle)*.

The result is minimal material stock in the process. This can function reliably only if the qualifications of the employee, the availability of the machines, and the intermediate products created in the process meet very high standards. All the previously mentioned initiatives and approaches for improving production, continuous maintenance of quality, etc. are part of the TPS.

5S Method

At the core of the 5S method is the creation of more functional, safer, and more pleasant workplaces. Efficiency, quality, order, and safety can be improved simultaneously using such a concept. Central to this is improving access to components and work materials, as well as identifying and implementing measures for the standardization and improvement of process security. For example, only the equipment that is actually needed can be found at the workplace. Every object is

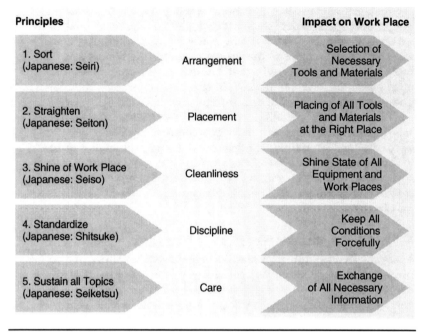

FIGURE 8.1 Principles of the 5S method.

assigned to a clearly defined place, and all operations are standardized. The individual measures are depicted in Fig. 8.1.

Wastage

The discovery and elimination of wastage are central components of the lean concept. In the Japanese approach, strictly minimizing wastage is a central issue. Wastage is anything that does not contribute directly to value creation. Every outlay for which the customer is not prepared to pay is seen as wastage.

This results in concentration on the value-creation process and classification into core process (i.e., creates direct customer benefit), supporting process (i.e., necessary for executing the core process), blind process (i.e., outlay without contributing to customer benefit), and loss process (i.e., destroys customer benefit already created). The latter two are to be avoided, and the former two should be organized as well as possible.

For production of articles, eight forms of wastage are often classified:

   1. *Overproduction.* All products, half-finished products, and services created without being requested by the customer. Most of the wastage types below are caused by overproduction as well as other factors.

2. *Inventory.* Inventories as production buffers cover weak points, but as overproduction, they tie up capital and space and create useless handling outlay. At the end, stocks often have to be written off and also appear in accounting procedures as a service that is not profitable.
3. *Transport.* Material transport does not bring any direct customer use to the product. Storage processes generally should be seen as blind processes.
4. *Waiting time.* Waiting times for processes, missing material, etc. tie up resources that are not used to create value in these periods.
5. *Overprocessing.* Through insufficient involvement of production in the development process, unsuitable equipment and unsuitable systems, etc., processes are difficult to monitor. This results in defects, reduces flexibility, and leads to error processes and waiting times.
6. *Long routes.* As in transporting material, long routes in production (from one production step to the next) are also inefficient.
7. *Defects.* Errors in products mean outlay for corrections (blind processes) or that performance is lost in scrapping (loss process). Furthermore, the process that was interrupted must run again (blind process).
8. *Unused potential.* All know-how and skills of employees in processes that are not used to improve the process are considered wastage.

We distinguish between avoidable and nonavoidable wastage. Many documentation processes, for example, cannot be avoided (this should be checked carefully) but are still wastage. Avoidable wastage should be eliminated. For example, employees can avoid making several trips to get tools using appropriate 5S measures.

6Sigma

6Sigma is a statistical quality goal (i.e., standard deviation based on zero deficiency) and is also the name of a quality management method. Its core element is the execution of data-oriented improvement projects through specially trained staff using tried and tested quality management techniques. Process improvement, reducing scattering, and achieving cost savings are the main goals of this method. 6Sigma is implemented today by numerous large companies not only in the production industry but also in the service sector. Many of these companies expect proof of 6Sigma quality in production processes from their suppliers.

The most often used 6Sigma method is known as the *DMAIC cycle* (*d*efine, *m*easure, *a*nalyze, *i*mprove, and *c*ontrol; see Sec. 2.4.6).

This is a project and control loop approach. The DMAIC core process is implemented in order to make already existing processes measurable and improve them continually. The larger the standard deviation is, the more probable it is that the tolerance limits will be exceeded. It also applies that the further away from the center of the tolerance range the average value is, the larger the percentage of exceeded limits will be. Therefore, it makes sense to measure the distance between the average value and the nearest tolerances in standard deviations.

The name *6Sigma* comes from the requirement that the nearest tolerance limit should be at least 6 standard deviations from the average value (i.e., 6σ level). Only if this requirement is fulfilled can a "zero-defect production" be achieved; that is, the tolerance limits are almost never exceeded. Altogether, there are seven different 6σ levels ranging from about 30 percent zero-defect production (level 1) to almost 100 percent zero-defect production (level 7).

Lean Production

The recognition and intentional avoidance of wastage and implementation of the 5S method are essential components of lean production/lean manufacturing. The concept was soon expanded by such ideas as lean administration and lean maintenance and extended to companies whose production was not characterized by large-scale serial or mass production and finally was developed further to lean management. This concept is a philosophy of excluding (down to the smallest detail) all unnecessary activities in production and in management through intelligent organization. It is based on innovative changes in the value-creation chain and the operators accompanying them.

8.2.3 MES for Reducing Sources of Loss

In order to decisively reduce the sources of loss described earlier (see Sec. 8.2.2), an integrated MES is needed in production. Where these sources of loss can be influenced considerably (i.e., eliminated or reduced) by the introduction of an MES, the benefit of the respective system is evident. Although these strategic initiatives are also supported by organizational measures, truly measurable benefits are brought about only by the IT-based support of an MES.

The individual benefits of an integrated MES are explained below. In particular, the benefits are examined where a return on investment (ROI) can be determined before implementation. Today, there are several scientific approaches to determining the ROI (see also [KAPLAN NORTON 1992]). However, these theoretical approaches cannot always be implemented in practice.

In the 1990s, the Manufacturing Enterprise Solutions Association (MESA) carried out surveys in more than 100 companies with different production types that had implemented an MES regarding the benefits of such a system. It should be noted that the extent and degree of maturity of MES have improved considerably in the meantime.

The companies surveyed considered the following benefits most important (this would hardly be different in the case of a resurvey of companies today):

- *Integrated data transparency*
- *Reduction of time consumption*
    - Administrative handling time
    - Planning time
    - Cycle time
    - Setup time
    - Input time
    - Waiting time
    - Transport time
    - Storage time
    - Overall processing time
- *Reduction of administration outlay*
    - Elimination, reduction of indirect value-creation activities
    - Elimination, reduction of documentation
- *Improved customer service*
    - Reliable delivery dates
    - Reliable information on order progress
- *Improved quality*
    - Automated proof of process capability
    - Reduction of scrap, reworks
- *Early warning system, real-time cost control*
- *Increase in personnel productivity*
- *Compliance with directives*

8.3 The Benefits of an MES

8.3.1 Integrated Data Transparency

What is lacking in today's production systems is the punctual recording of production data and their usually isolated evaluation using table calculation. Generally, there is no integrated overall picture of all data for an assessment of the overall situation. An MES is the instrument for integrated data recording and performance monitoring, on the one hand, in real time and, on the other, for more long-term analyses.

8.3.2 Reducing Time Usage

Savings are obvious when processing times for production orders are reduced. The time usage for processing an order can be divided into the following time areas:

- Administrative processing
- Operative order planning
- Setup
- Production
- Interim storage
- Final storage (Fig. 8.2)

Reducing Time Usage for the Administrative Processing of Orders

The time needed for the administrative processing of production orders can be reduced by direct management of the orders at level 3 = MES (see Sec. 3.2.1). Unnecessary administrative activities at level 4 = ERP can be avoided if the order data (i.e., article, quantity, and date) are forwarded immediately from level 4 to level 3. Thus the reliable determination of a delivery date can be expedited. The delivery dates transmitted are communicated directly to the customer or are made available to sales and marketing on short notice.

On the other hand, the general rationalization of administrative processes based on implementation of an MES is measurable; this is examined in more detail in Sec. 8.3.3.

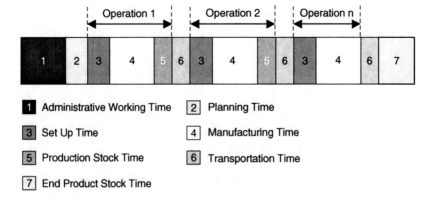

Figure 8.2 Time usage areas of an order.

Reducing Planning Time
Conventional planning systems, some of which are still used without electronic support, take up a lot of time and are not very effective. Planning data taken from ERP systems are seldom usable, and a considerable time outlay arises to balance load peaks at the planning control center. The result is planning that is not optimized and is incorrect, considering all process chains involved. The use of a planning algorithm that synchronizes all process chains with regard to the current availability of resources saves a great deal of time. Even a large order pool can be planned reliably in a few minutes. It is impossible to attempt to calculate these effects individually. It is often argued that savings can be verified by comparing the cycle times using conventional methods with the operative planning results of an MES.

Reducing Cycle Time
The respondents to a survey on MES [MESA 1997] stated that it was possible to reduce cycle times considerably by implementing an MES. It should be noted that this was certainly not achieved through MES but largely by implementing process measures or through a more rational organization of workplace handling (see Sec. 8.2.2).

In automated processes on a machine, this can be achieved by using new materials, new technologies, improved tools, etc. For example, in 1982, the cycle time for producing a CD was 22 seconds; today, it takes just 3 seconds.

Another measure for reducing cycle time can be achieved by arranging workplaces into work cells to minimize transport processes and manual handling (part of the TPS method; see Sec. 8.2.2).

The cycle time alone does not indicate the execution time. This is linked to distribution times, which, however, cannot be reduced by an MES through organizational measures.

Reducing Setup Times
With the planning methodology of an integrated MES, the sequence can be optimized with regard to cost aspects and minimizing setup times. This minimization is often accompanied by organizational measures such as the SMED (*s*ingle-*m*inute *e*xchange of *d*ie) method. SMED is an organizational method that tries systematically to reduce the switch time of product A to product B with a quantitatively achievable goal, for example, setup in less than 3 minutes. This makes it possible to respond quickly to changes on the market and to switch to new products. The reduced setup times affect the work plans and the setup matrix. Thus the cycle time is shorter per se in the individual work steps.

However, when reducing setup times, it is important to consider that in certain circumstances, saving time in setup means that storage times will be longer and therefore that storage costs will increase.

Here again, a detailed measurement of benefit will produce results that are hardly usable. Only the previously mentioned comparison of cycle times of the current system with cycle times using an MES will produce a calculable benefit. This should be carried out before the implementation of an MES.

Reducing Time Outlay for Manual Data Recording

In the MESA survey [MESA 1997], a large portion of the companies involved stated that the outlay for data recording could be reduced by over 50 percent by implementing an MES. Although it is possible to debate exact percentages, one goal of a qualified MES is to keep data-recording efforts as low as possible. Through the connection and integration of machines with the order process, measurement data are recorded automatically and saved in the MES. In manual processes such as the initiation of an order or operation, setup, and actual production, etc., the data recording is carried out mainly via an MES terminal (see Sec. 6.1.3), reducing input time considerably.

The allocation of material resources generally is carried out via material supply lists or accompanying cards with an identification code. If quantity counters are in use, it is no longer necessary to input quantities. Here, too, detailed measurement of benefits is difficult; only the comparison of the overall process can determine benefit.

Reducing Waiting Times

One source of loss in production companies is waiting times. This is generally due to the late supply of the required resources for individual operations. It most often affects raw materials and preliminary products.

The reason for this is a flawed planning system that is not capable of planning the orders in a synchronized, collision-free manner, taking into account the availability of resources. A planning tool integrated within an MES, as described in Sec. 4.2, largely can eliminate this source of loss or at least keep it as low as possible.

Reducing Storage Times

It is not possible to avoid storage times and therefore storage costs in the production process. However, by planning according to the pull principle, storage times and the associated storage costs can be reduced considerably. This affects raw materials, intermediate products in production warehouses, and the articles in final storage. In addition to the requirement-oriented supply of material (e.g., just-in-time, just-in-sequence, and E-*kanban*), the pull principle aims for fluid production (theoretically, one-piece lot size). However, there always will be at least small production interim storage if the cycle times are not harmonized with the individual machines (i.e., same cycle times). The MES with its qualified planning instrument is capable of reducing storage times for materials and articles significantly.

Reducing Transport Times
The reduction of transport times is achieved largely through conventional measures in the logistics sector. The MES can contribute if transport activities are integrated into the process chain of the order as an operation with planning times.

Overall Processing Time
In the points just listed, a possible reduction of time usage for individual orders was discussed. It is very difficult to calculate benefits precisely by reducing the individual time types. However, there is an approach that allows time savings to be calculated relatively precisely in their entirety before the implementation of an MES:

1. The cycle times of completed orders exist in a conventional system. A representative selection of articles is taken, and the processing times for individual orders are recorded for a statistically representative period.

2. Next, the article sequence scenarios are summarized for these articles. Using a simulator, the master data are mapped with the work plans. The processing of the orders then is simulated with the planning tools. For realistic values, a deviation value is added to the planning results.

3. The processing times measured are compared with the processing times of the conventional system. Simulations have shown that savings averaging 30 percent can be achieved. In order to make the results more concrete, as many scenarios as possible for the articles concerned should be simulated. An average reduction of 30 percent also was recorded in a survey of U.S. companies [MESA 1997].

This approach makes it possible to prove the huge usefulness of an MES just from the point of view of processing times. The other benefits associated with implementation of an MES are not even considered.

8.3.3 Reducing Administration Expenses
Generally, there is considerable savings potential within administrative activities. The aim is to check to what extent indirect value-creation activities can be eliminated or reduced considerably through the implementation of an MES.

The following method, which was used at the peak times for rationalizing administration (i.e., 1970s and 1980s), should be used for this purpose:

- Listing the activity profiles of employees in administration
- Time- and quantity-related weighting of the activities by employees

- Listing of a task and time distribution plan
- Analysis of every activity with regard to the necessity or potential for improvement
- Creation of a rational workflow handling integrating an MES

Analyses in the field of administration rationalization led to the conclusion that with conventional methods alone, savings of at least 20 percent can be made. If you couple the analyses with the effects of an MES, the savings effect certainly will rise to over 30 percent. Indirect value-creation activities are eliminated or at least reduced considerably.

Below is an example from the MESA survey [MESA 1997]:

- Over 60 percent of respondents indicated that the number of documents used could be reduced by more than 50 percent by implementing an MES. With a good planning tool within the MES, the personnel required could be reduced by at least 20 percent.
- In an individual case, it was demonstrated that 6 of 20 jobs in administration could be cut though conventional rationalization and implementation of an MES.

8.3.4 Improved Customer Service

Reliable delivery dates and information about order progress are essential on the market today. With an MES, these requirements can be suitably covered. Although the benefit of these options cannot be measured directly, the image of a production company improves, and order increases are more probable.

8.3.5 Improved Quality

The quality assurance systems used in the past were mainly isolated systems that were used largely independently of the other recording and monitoring systems. The frequently mentioned norms [ISO 9000] were the main factors driving the implementation of these systems. Today's norms are oriented toward all elements of product quality. The entire production process should be viewed in an integrated manner and documented accordingly in order to supply the customer in particular with the highest possible product quality and be able to trace this in all aspects at any time. Here, the integrated view of control of process [statistical process control (SPC)] and quality [statistical quality control (SQC)] should ensure and decisively support the objective of zero-defect production (as Motorola formulated it in its DMAIC concept in the 1980s).

Within the scope of the strategic 6Sigma initiative, an MES becomes a significant tool. The ensured process capability is largely supported

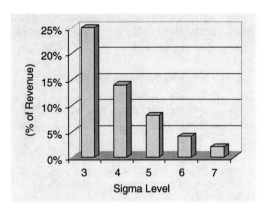

FIGURE 8.3 Sigma level and associated error costs based on poor quality.

through continual, automatic measurement and monitoring of process parameters on machines in an integrated MES. The MES supplies the data for the DMAIC method in connection with actual production. In this closed-loop method (see Sec. 8.2.2), a continual improvement of product quality is attempted. The data of the production process are central to this method. It is not possible to measure the impact of an integrated MES on reducing rejects and reworking in advance. If the tools mentioned are used correctly (this also requires correspondingly qualified personnel), it should be possible to reduce rejects and reworks considerably and thus also reduce the costs associated with this source of loss. In the MESA survey, respondents confirmed an average reduction of 15 percent in rejects and reworks [MESA 1997].

Another aspect of the 6Sigma initiative is the correlation between the sigma level and error costs associated with it. The lower the sigma level, the higher the error costs will be. Most production companies today are located in the range of 3σ to 4σ. An investigation has revealed that error costs make up about 25 percent of turnover at a 3σ level, and at a level of 4σ, losses still account for 15 percent [ZARNEKOW BRENNER PILGRAM 2005] (Fig. 8.3).

8.3.6 Early Warning System, Real-Time Cost Control

Unacceptable deviations are recognized immediately by the real-time control of all influencing parameters in a production process, and measures can be taken accordingly. Primarily, real-time cost control acts as an early warning system in connection with the reasons for exceeding planned costs.

8.3.7 Increasing Employee Productivity

An integrated MES provides machine workers electronically with real-time information needed for orderly production with as few errors as possible. It is also impossible to quantify this benefit, but its benefits can be seen clearly. The MESA survey [MESA 1997] revealed that most companies value this aspect very highly.

8.3.8 Compliance with Directives

Several institutions have drawn up directives, often in the form of regulations. The aim is always to ensure the best possible production process with the highest level of quality and to document this. The Deutsches Institut für Normung (DIN) standardization body for quality assurance has defined requirements that companies must comply with (see Sec. 3.2). The Food and Drug Administration (FDA), with its guidelines on good manufacturing practices (GMPs) and 21 CFR Part 11, sets the standard for chemical products and products in the food industry (see Sec. 3.2.4). Therefore, it must be the goal of a company to comply with these regulations, become certified, and have Electronic Data Processing (EDP) systems be validated.

An integrated MES with its recording, control, and documentation functions supports these requirements to a considerable extent. If companies use such integrated systems, they are rated correspondingly higher during audits.

8.4 The Costs of an MES

The costs of an MES, like those of any comparable complex IT system, can be divided into the following areas:

Consultation Costs before Implementation

Before acquisition of an MES, it is recommended that you investigate the benefits of an MES individually, specifying the required fields and requirements of the future MES in order to avoid making a wrong decision. This is best ensured using competent, neutral consultants.

Acquisition Costs

These are all costs that arise at the beginning of a project that are associated with the actual purchase of the system. These include hardware, system software, application software, personnel (if additional personnel are needed), and consultant costs.

Adjustment Costs

Compared with office products, the software delivered by suppliers can seldom be used in the form in which it was developed. Thus adjustment and expansion costs arise. Unfortunately, this outlay is very high for many of the systems on the market at the moment because the systems offered fulfill the requirements described only to a limited extent.

Implementation Costs

Every system, even if it is supposedly "intuitive," gives rise to introduction and training costs.

Running Costs
These include all running costs that are caused by using the system. Maintenance and support costs are typical examples.

8.5 Summary

At the start of this chapter we examined the measures and strategies available in the corporate environment to determine the cost-effectiveness of investments. The explanations showed that these methods usually are not suitable for assessing the cost-effectiveness when implementing an MES.

Later, existing measures for the specifically reducing sources of loss in production were explained. These measures are mainly possible only through support of a suitable IT system. As can be seen in the examples, an integrated MES will bring immense benefits to a company with regard to reducing the sources of loss shown.

It is very difficult to measure these specifically before implementing the system, and it is only partly possible. If we look at the individual benefits together ("An entire system is always more than the sum of its individual components."), these are so great that discussions about the cost-effectiveness of an MES are unnecessary as long as the boundary conditions mentioned are present. Professional project implementation is especially important. The ROI that can be expected after implementation is generally 2 to 3 years from a TCO point of view.

CHAPTER 9
Implementing an MES in Production

9.1 Implementing IT Systems in General

9.1.1 Selection of Components

If a new information system is to be installed in a company, the technical components, software and hardware, must be procured first of all. Then the implementation of the technical components in the already existing information technology (IT) landscape of the business follows. Training employees and adapting business processes to the information system are associated with this aspect.

The *selection* of the hardware components depends on the software. Here, it must be decided whether an individual computer or a computer network should be procured, what processor architecture is needed, and what other components are necessary. The selection of software is generally much more difficult than the selection of the hardware components.

If new software is needed, the question then arises whether standard software should be bought or individual software developed. *Standard software* is already preproduced and covers one or more business processes completely with one or more programs. *Customized software*, on the other hand, is created especially for an organization and can be developed by either the organization itself or an external supplier. The advantages and disadvantages that generally exist in practice for both standard and customized software are listed in Table 9.1. In practice, a mixed variant of standard software with a small number of individual extensions is found.

If customized software is to be developed, it must be checked whether this can be developed within the company or if a software development company should be engaged to develop the software. If one or several standard software products correspond to the requirements of the customer, one of these products should be selected.

Standard Software	Customized Software
Is often less expensive than individual software	Software is tailored precisely to the needs of the organization
Support provided by software manufacturer	Implementation normally takes place in stages and without adjustment outlay
Software has already proved itself in use	It is likely that no interface problems will arise
Extensive documentation generally is provided	Only desired functions are implemented
Software available immediately	Developing the software can lead to unforeseeable costs
Adjustment to individual needs is necessary (higher adjustment outlay)	Quality of the software is reduced by an inexperienced development team and time pressure
Interface problems could potentially arise	Documentation is often neglected
It is possible that not all requirements will be met	All requirements are met
Functions that are not required must be purchased with the software	Only the required functions are paid for

TABLE 9.1 Advantages and Disadvantages of Standard and Individual Software

We need to distinguish between general and software-based criteria when it comes to the selection. The general criteria serve to evaluate the manufacturer and the order. The manufacturer should be assessed based on references and its own information with regard to the provision of maintenance and services, the costs for software and other services, as well as its economic situation.

The software itself can be assessed based on the following criteria:

- *Functionality.* The software must be capable of fulfilling all functional requirements.
- *Quality.* The software should contain as few errors as possible and be able to cope with input errors.
- *Performance.* The functions should be carried out not only correctly but also within an appropriate time frame and with reasonable resource requirements (e.g., main memory and processor load).

- *Documentation.* The users should be supported by the software in every situation by means of corresponding documentation.
- *Technology.* In order to offer the option of later maintenance and extension, the software should not be based on old technologies (e.g., programming language, programming concept, etc.).

9.1.2 Implementation Strategies

Once suitable software is selected or corresponding individual software is developed for the requirements of the business, it is possible to continue with the *implementation* of the information system. Systematic implementation is unavoidable because of the complexity of information systems. This generally can be carried out on the basis of three strategies:

- "Big bang"
- Step-by-step implementation in individual operating areas
- Step-by-step replacement of individual business processes

The *"big bang" strategy* foresees the installation of an information system in one piece. Here, all affected business processes are handled via the new system by a certain deadline. This leads to a high risk that errors in the information system could affect the entire organization. Errors can be caused not only by the software and hardware but also by the people handling the technical components. In order to avoid user errors, an extensive training of all users should be carried out almost in parallel because they normally will begin to use the system at the same time.

In order to reduce the risk of an error, it is possible to implement the information system in individual *operating areas* gradually. In this approach, initially the business processes for only one division are supported by the system. The advantage is that possible errors do not affect the entire business, and thus the users can be trained gradually. Experience gained with the information system in the parts of the business in which the system is already in use therefore can be transferred to the other parts.

Step-by-step implementation also can be carried out with regard to the *business processes.* Here, only some processes are carried out initially via the new system. The risk of failure therefore is greatly reduced and becomes manageable. User training also can be done in stages. This strategy is characterized by lower risks but a higher outlay in terms of time. Selecting the suitable implementation strategy therefore is somewhere between the two generally competing goals of risk reduction and outlay minimization.

For the *acceptance of new information systems,* intensive preparatory training of users is just as important as careful instruction of the employees who will be affected either directly or indirectly. Motivating effects to create qualified advocates for the system implementation occur only when the training is based not on merely practicing user maneuvers but on when it affects the entire task context and thus makes benefits apparent for every user. Skepticism always arises among individual employees when obvious development flaws are not corrected successfully. It also has emerged that lack of information leads to loss of trust and makes employees uncertain. In addition, it is especially difficult for older employees to accept new systems.

For complex or new tasks such as implementing an information system, it is helpful to differentiate between the phases of requirement analysis, development, procurement, and implementation. A work group is very suitable for this purpose, whereby the project manager's competence should be tightly limited and extend to management tasks only. It often can be seen that the decision for innovation meets up against barriers of will or skill. Barriers of will can be explained mainly by the desire to stick to the known and trusted. They are often difficult to overcome, and doing so generally requires the supervisor's position as well as specific motivational steps. Barriers of skill, on the other hand, can be overcome by know-how. When information and communication systems are introduced, both power and expert know-how are important.

9.1.3 Problems during Implementation

Among the problems that go hand in hand with the implementation of information systems, it is possible to differentiate cognitive problems, changes to working conditions, and changes to working requirements.

Cognitive problems generally can be traced to selective information absorption and processing. Information is perceived and processed in a subjective, structure-oriented manner. From this standpoint, the preference for hard information over soft information is especially important for decisions about technological investments. *Hard information* denotes all concrete, easily comprehensible and traceable information. This problem can be alleviated at least in part through a corresponding organization of information perception and processing.

However, this does not mean that *soft information* is generally to be neglected or presents a problem. For decision types such as strategic decisions, personnel decisions, research and development, etc., soft information is at the center of interest. For example, only very general, global contexts are recorded at the strategic planning level. According to the tasks of strategic planning and the degree of detail of this information, only general goals, general procedure guidelines, etc. can be derived from them that are then made more precise and "hardened" at the next planning level. Known cognitive problems for technological investment decisions are, for example, the neutralization of subsequent costs, (non-) perceived responsibility, ability of the manufacturer to adjust, and what is known as *gaming*.

The discussion about *changes in working conditions* has been hotly debated since the beginning of the commercial use of IT. In particular, there are quite contradictory statements. They mainly concern questions of how (and if) job descriptions and the frequency of social contacts will change. In both cases, enquiries are made about changes with regard to appreciation and depreciation of tasks and abilities.

The correlations between the use of technology and working conditions are essential for labor organization. Numerous assumptions and speculatively formed hypotheses are based on these correlations. These assumptions and hypotheses (e.g., reduction of social contact and reduction of decision-making scope) have been proved or disproved in various empirical investigations that have since taken place. This situation, which initially appears to be contradictory, however, leads inevitably to the conclusion that implementation of a new information system leads to a great deal of organizational leeway. In addition to technological conditions, company goals and the needs of the users in particular should be taken into account. It is clear that many people find the reduction of monotonous routine tasks a relief and that they can then apply themselves more to the qualitative aspect of their work. Thus there is more time for solving less structured problems, for coordinating with other employees, etc. Opponents counter these arguments in favor of the system by stating that most of the tasks and decisions are structured and therefore can be programmed. This can lead subsequently to a balancing of tasks between group leaders and their employees.

Relevant studies have already shown that rationalization measures and technology applications conceal various risks for most of the employees affected, which contributes to the deterioration of working conditions and business relationships. Known risk areas are as follows:

- Job security
- Work intensity
- Qualification
- Earnings
- Working conditions
- Labor relations
- Working time regulations

Changes with regard to work requirements affect, among other things, competence, qualification, work content, and work intensity:

- *Competency.* This refers to "the perception of learning opportunities and the associated ability to go beyond the limits of one's own workplace." Generally, it has been seen that the increase in competency is dispersed very widely through

IT-based work and can be found in almost all employment groups.

- *Qualification.* New information systems can create feelings of anxiety and stress. For example, anxiety arises if employees fear that they will not be able to cope with the new technology or that they will not achieve the necessary qualification. Overloading resulting in this situation can be combated by means of various measures, such as early information for employees affected with regard to planned actions and their expected effects, timely and thorough training, and good instructions or a corresponding offer of retraining or further training activities. Determining the actual qualification level is difficult owing to existing conflicts of interest. Employee representatives will be interested in setting qualifications at a high level and allowing this increase to be reflected in the wage agreement. On the other hand, companies instead may attempt to argue in favor of computer-supported workplaces with a low level of qualification. Automated work in this way becomes a job for "semiskilled workers."

- *Work content and work intensity.* Unlike qualification, studies in the past have shown that IT implementation in the office, for example, has had hardly any effect on work content. Where changes were mentioned, however, they were mainly deteriorations. However, change depends on the main function of the activity and on the branch of the economy concerned. The most significant worsening was determined in businesses with mainly input and user activity. On the other hand, in accordance with results from other comparable research, consultation, management, and organizational activities were least affected by deterioration of work content. In summary, we can state that hopes for enriching work content are seldom fulfilled. The work relief that follows implementation of an information system is usually filled immediately with additional activities. Thus work relief is transformed into a reduction of costs and rationalization and then often results in increasing work intensity.

9.2 Preparation of the Implementation Project

9.2.1 Establishing the Core Team

The implementation of a manufacturing execution system (MES) is a complex project and therefore requires strong project management. The MES is increasingly becoming a strategic decision-making tool for management. Therefore, it is necessary for the business management/board of directors to stand behind this product and give all necessary support to all those concerned. Since implementation of an MES affects all aspects of production, all departments affected also must be represented by spokespersons on the core team. As a rule, these departments are work preparation, production, logistics, quality assurance, maintenance, internal IT, controlling, and especially the *company management*. The representatives of the departments must receive sufficient decision-making competence and also a defined time allowance for their additional activities on the core team. The core team is often represented as a *control circuit* because one of its main tasks is primary control of the project.

The core team should carry out the following activities in the implementation project:

- Project management throughout the entire project period
- Reaching the fundamental decision: "MES: yes or no"
- Appointing a project manager and project team for the actual implementation
- Providing financial resources
- Releasing employees for duties during implementation
- Monitoring project progress using milestones, including cost and deadline monitoring
- Review and success monitoring after implementation
- Creating an operating concept for the sustainable use of the system

9.2.2 The Fundamental Decision: MES: Yes or No

In order to limit the time and cost outlay for system implementation, a *phase model* is, for example, a suitable solution, in which it is possible to stop the entire project after certain phases. The first phase includes a fundamental analysis of the situation within the company. At the end of this analysis is the decision "MES: yes or no." This fundamental decision and all subsequent decision-making processes throughout the course of implementation should be prepared and documented on the basis of reliable criteria. The decision matrix in Fig. 9.1 with criteria can, for example, facilitate the fundamental decision—"MES: yes or no."

9.2.3 Establishing the Project Team

If the fundamental decision is made to implement an MES, the next step is to establish the project team with a *project manager*. Whether a project team is required in addition to the aforementioned core team depends on the complexity and duration of implementation. The proportion of *internal services* (services that are provided not by the system

Criteria / Importance:	Low	Medium	High
Process Operations	<5	5–10	>10
Products	<10	10–50	>50
Self-made Sub-assemblies	0	1–10	>10
Product Versions	0	1–10	>10
Added Value in Production	Low	Medium	High
Information Requests per Month	<20	20–100	>100
MES Required?	No	Recommended	Yes

FIGURE 9.1 Decision matrix for the fundamental decision: "Is an MES necessary?"

supplier but by the company itself during implementation) also can be a criterion. In any case, a project manager is required to plan and monitor the project progress, to act as an interface with the system supplier, to include all internal departments affected, and to report on the overall progress in the core team. This ensures that the project is embedded in the company organization and that preparation of the actual project is concluded (Fig. 9.2).

9.3   Analysis of the Actual Situation

9.3.1   Introduction

Before implementation of a (new) MES, it is absolutely essential to analyze the actual situation in detail. Only on the basis of this analysis can potential for improvement be discovered and thus requirements be defined for the new system. Analysis of the technical boundary conditions is also important to avoid unpleasant surprises in terms of project costs. The *main focus* of the analysis must be on the *work processes in production*. From these, the functional requirements

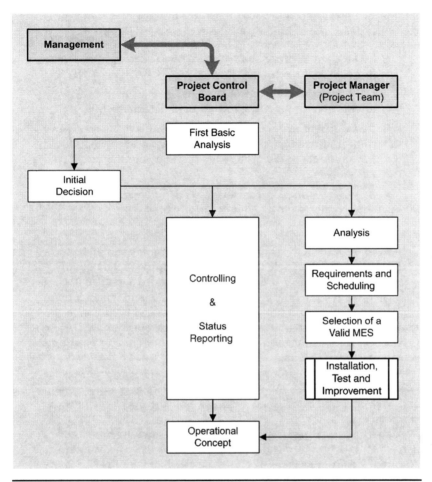

FIGURE 9.2 Overview of project progress and implementation within the organization.

for the MES must be derived, and in certain circumstances, organizational changes in the processes must be implemented within the processes themselves. Finally, the success of the project must be proven for controlling and company management on the basis of measurable facts. For this purpose, key figures should be defined and determined in advance. Precisely these key figures then can be compiled continuously during the implementation phase of the (new) MES, thus making success of the implementation transparent for all those involved.

9.3.2 Existing Infrastructure

The *integration of control systems* from production offer, as shown in Sec. 7.3.2, considerable advantages over an exclusively manual data-recording system. However, to be able to connect to the existing control systems, the technical conditions must be checked:

- Do the production-control systems have a suitable physical interface with the network (ideally, Ethernet+TCP/IP-enabled)?
- Do the production-control systems have a suitable software interface for transferring the required data to the MES?
- Do the machines/systems have visualizations or supervisory control and data acquisition (SCADA) systems that could transmit data to the MES as an alternative to the actual control systems?
- Are the production-control systems and visualization systems connected to a network (ideally Ethernet)?

If these conditions are not present, high costs may arise for the desired connection of the control systems. This outlay should be compared with the added value that arises with improved data quality.

In the second stage, it should be investigated whether *isolated solutions* already exist for different *partial functions* and whether these isolated solutions can continue to be used and need to be connected to the MES. Examples here are machine data acquisition (MDA) or production data acquisition (PDA) systems that already exist and provide satisfactory data.

A further important aspect is the *connection to the enterprise resource planning (ERP) system*. If this exists, the task split between ERP system and MES must be established precisely, and the interfaces must be specified exactly.

Finally, the existing *IT infrastructure* must be investigated to establish which interfaces are still necessary for the MES. The question of "desired" database systems also must be clarified with the IT department. It is possible that there is still capacity available for the MES in an already existing database server.

Direct requirements for the MES can be derived from this review of technical conditions and then be integrated into the contract specifications.

9.3.3 Existing Processes and Required Functions

Investigation of the existing processes is a condition for discovering potential improvements. External consultants are better suited for a truly objective investigation than employees of the company, who already have a particular "angle." One possible tool for this investigation is *value-stream mapping*. Value-stream mapping is a method used in lean production (see Sec. 8.2.2) that maps the entire material and information flow (separately for every product) of the value stream from the end customer to the supplier of the raw material.

Function	Actual Manual Function	Actual Isolated Solution	MES Integration
Fine planning and control center			
Operating data capture			
Machine data capture			
Material management			
Tracking and tracing			
Quality management			
Cost control			
Performance analysis (KPIs)			
Compliance management			

TABLE 9.2 Required and Existing Partial Functions of the MES

As a result, value-stream mapping provides the "real" processing time (sum of the value-creation processes) and the overall cycle time of a product. Non-value-creation processes [in the sense of the Toyota production system (TPS), "wastage"] thus should be recognizable.

However, functional requirements for the MES also can be derived from the results, and for example, optimized fine planning can be used to reduce production bins and minimize cycle times. The core functions of the MES that are actually required are listed in Table 9.2. If these functions are already present in existing IT systems as *isolated solutions* (see Sec. 9.3.2), their integration into the MES can be evaluated.

9.3.4 Key Figures as the Basis for Monitoring Success

Goal of Investigation
Implementation of the MES generates costs and therefore must also lead to measurable improvements in production. Reliable success monitoring is best done using reliable indicators. Thus key figures must be found that describe the current situation (before introduction of the MES) and should be improved by the MES, thus reflecting an increase in effectiveness.

Quality and Up-To-Date Status of Data
The quality of the existing data that are used to calculate the key figures must be checked to ensure that they are up to date and correct.

This applies to any data that have been gathered automatically as well as to "manual recording," such as reports of disruptions in shift logs, initial quantities of the raw materials used, or reworked parts at manual workplaces. The quality of these data can be established by means of targeted random tests. It also can be useful to set a defined period of time for gathering data (e.g., a week) and then use the data from this period as a benchmark (basis data for a comparative performance test).

Exact Definition of the Viewpoint: Avoiding Moving Targets
If the quality of the basic data has been ensured, calculation of the selected key figures can go ahead as the next step. It is absolutely necessary to reach a consensus about the formulas that will be used to calculate these key figures in advance. A success-monitoring method can provide reliable results only if the calculation methods are not changed when the MES is implemented; the notorious comparison of "apples and oranges" must be avoided.

9.3.5   Suitable Key Figures for Success Monitoring

Overview
The key figures selected should be reliable with regard to the general level of quality in the company and should, of course, lie within the area of influence of the MES. The key figures collected are an excerpt from a large number of parameters that can describe a company in qualitative terms. Here, it is important not to collect too many parameters but rather limit the choice to a small number of reliable parameters. Generally, there are also well-known problem areas whose causes should be alleviated by implementation of the MES. Then it is also worth looking at these problem areas in particular.

Indirect Value Creation
The share of indirect value-creation activities in the entire value-creation process is a measure of the leanness of the production, or a *lean production benchmark*. Indirect value creation includes marketing, purchasing, controlling, and work preparation. However, non-value-creating activities also can exist directly in production, caused by poor process organization or logistics.

The percentage of indirect value creation can be defined as follows as a percentage (value creation = VC):

$$\text{Indirect VC} = \frac{\text{number of employees from indirect VC}}{\text{number of employees from total VC}} \times 100$$

The proportion measured is a first indication of possible savings (e.g., personnel, documentation, etc.). This can be supported by using the reduction of indirect value-creation method explained in Sec. 8.3.3.

Order Cycle Times

One important goal is reduction of the average cycle times for orders (i.e., products). To determine the highest potential, the current cycle times for a typical product must be recorded. For this product, a simulation then is carried out with the aid of the fine-planning component integrated within the MES. The outlay that arises through parameterization of the system to map out the current production sometimes can be relatively high. On the other hand, this simulation also provides a real outlook with regard to potential savings that implementation of an MES makes possible.

The potential for improvement with regard to cycle times (CT) can be measured as a percentage as follows:

$$\text{Improvement potential for CT} = \frac{\text{CT\_actual} - \text{CT\_simulation}}{\text{CT\_actual}} \times 100$$

Failures/Reworks

Failure and rework quotas are an indicator of the level of quality of production. By using a statistical process control/statistical quality control (SPC/SQC) tool set within an MES and implementing the 6Sigma projects derived using the define, measure, analyze, improve, and control (DMAIC) method (see Sec. 8.2.2), these quotas can be improved. The measurement of the actual condition can be carried out based on the following formula. The target condition that can be achieved through implementation of an MES must be estimated in some cases or established arbitrarily as a target value:

$$\text{Failure quota} = \frac{\text{quantity\_failure}}{\text{quantity\_total}} \times 100$$

OEE Key Figure

The most important and most widely used key figure for assessing the efficiency of the existing machines and production equipment is *overall equipment efficiency* (OEE). The OEE is the product of three other key figures, each of which depicts a different aspect of the performance of a machine:

- *Availability = AV (in percent)*. Availability indicates for what period of time with regard to a defined total period (known as the *planned bin occupancy period*) the machine is actually ready for production, or available. The proportion of losses (i.e., machine not ready for production) includes stoppages owing to malfunctions or setup times.

- *Performance rate = PR (in percent)*. The performance rate is defined as the ratio between the quantities actually produced per unit of time (e.g., pieces per hour) and the theoretically possible quantity per unit of time. Thus it includes idle times and exceeding planned cycle and processing times for a machine/equipment.

- *Quality rate = QR (in percent)*. The quality rate is defined as the ratio between the *usable* quantity produced and the overall quantity produced. Thus it includes production units that are lost owing to defects or rework.

From this, it follows:

$$OEE = AV \cdot PR \cdot QR$$

This key figure should be recorded for defined and unchanging periods of time (e.g., shifts, days, and weeks) in order to make changes in the situation apparent immediately.

Inventory for Material Warehouse

Another key figure for the effectiveness of production is the *inventory turnover* in the material warehouse and the related warehousing times. The inventory turnover indicates how often the average value of the inventory is turned over per year. In running production with requirement-oriented supply, the material can be warehoused only for a short time or is delivered directly from the production line.

$$\text{Storage turnover} = \frac{\text{annual turnover}}{(\text{inventory value at beginning of year} + \text{inventory value at year end})/2}$$

$$\text{Storage time} = \frac{365}{\text{storage turnover}} \quad \text{(storage time in days)}$$

An MES with an integrated material management and operative planning function can increase inventory turnover and reduce warehousing times. The savings from tying up low amounts of capital that result from a higher inventory turnover must be contrasted with the risk of production stops owing to a lack of material.

9.3.6 Other Factors for Success

Overview
These key figures are measurable and thus verifiable factors and are therefore especially suitable for goal monitoring. In addition, there are many soft factors that cannot be measured without more input but that still represent a significant proportion of the company's success. Some of these factors are listed below.

Customer Satisfaction

Customer satisfaction depends on many factors. Naturally, products must be of a high quality and also must represent good value for money. However, reliable delivery dates, fast responses to inquiries, short-term changes to the scope of the order, or the option for information on the progress of an order also can be decisive for customer satisfaction. These factors can be improved when an MES is implemented. In some cases, it is necessary to check whether these factors are especially important for the customer satisfaction of the company and therefore should be included in the focus of the implementation project.

Employee Motivation

Value creation and quality also depend greatly on the motivation of the workers even in highly automated production. An MES can contribute significantly to improving motivation through reliable planning and transparent representation (through graphs) of the current situation. Achieving these targets also can be supported using key figures such as average information time or overall time of a status report from production control to display medium.

Cost Control

The necessity of real-time cost control should be checked in individual cases. This need has increased owing to cost pressure because real-time cost control can affect company revenue at short notice within an early-warning system. *Cost control* here means monitoring all costs (direct and indirect). A qualified MES offers real-time cost control at the operations level.

Order Tracing

Functional complaint management (another factor that influences customer satisfaction) and in many cases also legal requirements with regard to liability and guarantees make it necessary to be able to trace an order in its entirety. Questions such as "In what products were parts from batch *xy* used?" must not take days or weeks to answer; it is not sufficient to save the data "somehow"—fast responses are necessary.

Support for Strategic Initiatives

Today, there should be and are a range of strategic initiatives in production companies that aim at running production as cost-effectively as possible. These are mainly the Toyota production system (TPS), lean production, and 6Sigma approaches (see Sec. 8.2). An MES should be an accompanying tool for implementation of these initiatives.

9.4 Creation of a Project Plan

After the requirements and goals of the project have been established, a project plan must be drawn up that contains the main deadlines and the availability of internal resources. In addition to the project manager, other employees also must reserve time for the project. Implementing the system is primarily the company's responsibility—the system supplier can achieve the goals established only with the company's massive support. Above all, the challenges in the areas of employee qualification and acceptance can be handled from outside the company only with great difficulty.

The following list containing phases and milestones can be a rough guideline and must be tailored to suit the needs of the project. All required points must be included in the project plan with a clear classification of responsibilities and resources:

- Compiling contract specifications and invitations for tenders
- Assessing the tenders and allocating orders
- Compiling specifications based on the contract specifications
- Strengthening the internal infrastructure (e.g., network, production controls, etc.)
- Acquiring and supplying hardware systems
- Instituting project-specific adjustments to the system (customizing)
- Delivering and installing the software components
- Executing at least one pilot test
- Parameterizing the system
- Achieving parallel operation with existing systems in some sections
- Instituting function test(s) and optimization per production section
- Training measures for employees affected
- Approving the system
- Creating an operating concept
- Handing over operations to those responsible

9.5 Contract Specifications

The *contract specifications* form the basis for inviting tenders and contain the scope and goals developed in the course of the analysis phase. The following suggested structure is based on the scope described in Secs. 9.3 and 9.4:

- Description of the actual situation and infrastructure (see Sec. 9.3.2)
- General goals (see Sec. 9.3.6)
- Quantitative goals (see Sec. 9.3.5)
- Functional requirements (see Sec. 9.3.3)
- Quantity structure for machines, articles, number of terminals, number of users, etc.

- Interfaces with technical and content-related description
- Project plan (see Sec. 9.4)
- Description of the supply and performance scope (see Sec. 9.4)
- Appendix with delivery conditions, applicable norms, etc.

9.6 Selection of a Suitable System

9.6.1 Market Situation

From the user's point of view, it is very difficult to gain an overview of the market. *MES* has become a buzzword that is interpreted differently by every user depending on his or her viewpoint. In many cases, existing products that mapped a particular part of an MES simply have been renamed in MES. This often affects PDA and MDA systems as well as products for quality assurance. New functions were added to these basic systems through in-house development or subsequent spending. However, the resulting patchwork systems have no all-encompassing concept and design. Some suppliers of ERP systems simply dispute the need for an MES. These suppliers have the vision that an ERP system can take on all tasks from production management and data recording. This means that the MES functions are completely integrated into the ERP system.

For the potential user, the situation therefore is difficult but not hopeless. In the past few years, some systems have arisen among the ERP systems with MES sections and patchwork systems that rightly carry the name *MES*. It is worthwhile to find the system that is best suited to the needs of the company and is based on modern technology with secure prospects for the future.

9.6.2 Short-Listing and Limiting to Two or Three Applicants

A short list of suppliers and then a final decision on one system supplier should be made by the core team (see Sec. 9.2.1) plus at least one competent worker/system manager and the project manager. Workers and equipment operators should be integrated at this early stage in the project to ensure acceptance of the selected supplier/system. Which companies should be invited to submit a tender? If the core team has been examining the topic for some time, some potential suppliers probably are known already. Other applicants can be suggested by the purchasing division. The supplier should not be located too far from the installation site. Otherwise, especially for smaller projects, the ratio of transport costs to the overall costs is not economical.

The actual short list can, for example, be made using a list of questions provided to each supplier along with the requirement specifications. This list must contain all technical and functional requirements from the contract specifications and also a range of general questions on the supplier and the system, as shown in Table 9.3.

Chapter Nine

No.	Requirement/Question	Present	Present in Part	Not Present	Reply/Comment
101	Control center with Gantt plan				
102	Operative fine planning with . . .				
103	Resource optimization				
104	Setup time optimization				
.	.				
.	.				
.	.				
201	MDC automatically via interface to production				
202	MDC via machine terminal				
203	Performance analysis (KPIs)				
204	Fixed key figures, own key figures (formulas) can be defined				
.	.				
.	.				
.	.				
501	Database system				
502	Platforms for server modules				
503	Technology for client				
.	.				
.	.				
.	.				
901	Number of installations in a similar environment				
902	Average number of machines				
903	Average number of terminals				
904	Average number of articles				
.	.				
.	.				
.	.				

TABLE 9.3 Example of a Requirements Summary for Short-Listing Potential System Providers

After being included in the offer, requirements still can be complemented by estimates with regard to the supplier's situation (e.g., size, stability, delivery performance, delivery reliability, transport time, etc.) and the system (e.g., one-stop shop, modules purchased, etc.). Making an assessment of the requirements that is as objective as possible is the most difficult task in the selection process. An initial estimate can be made on the basis of knockout (K-O) criteria. For example, if detailed batch tracking from the end product to the raw material is a core requirement, other points cannot compensate for the lack of this function—such suppliers are therefore eliminated.

Cost monitoring should not be in the foreground in this selection phase. However, if the costs contained in the tender exceed the established budget by a large amount, this also can be a K-O criterion.

If an additional limitation is necessary, the individual points of the requirements should be weighted using a points system. Establishing the points (e.g., from 1 to 5) and the K-O criteria always must be done within the team and before the tenders are received in order to maintain objectivity.

9.6.3 Detailed Analysis of the Favorites and Decision

General Consideration

The suppliers that remain after short-listing them should be examined in more detail. If the outlay is manageable and also supported by the provider, a test installation is the best way to really get to know the system. It also can be worthwhile to visit a business in which the system is already installed.

The people who are to work with the system, that is, workers, equipment and machine operators, maintenance technicians, etc., must be involved in the decision-making process. Questions of user friendliness and suitability in practice can be judged best by these employees. The core team must collect and document their evaluations. The decision-making process must be transparent and traceable for all. Later discussions such as "If we had only . . ." thus can be avoided.

Even with the best preparation for the decision and the most detailed documentation of the arguments, there still can be opponents of the system. For the continued progress of the project, there are two alternatives for dealing with this circle of individuals:

- Try to convince these employees of the advantages of the decision, although this can be difficult or even impossible. Especially when old systems are to be replaced or work processes completely changed, tenacity often can be astonishing. To compensate for the "loss" suffered by these employees, only arguments as to real benefits can help. *Benefits* refer not to the benefits of the system for the company but to the benefits for the individual workers in their daily work. These can include points such as shorter pathways through installation of additional terminals, easier use and therefore time savings in order responses, less work and setup efforts through optimized sequence planning, etc.

- If, despite the aforementioned efforts, you do not succeed in convincing employees, these people should not necessarily be involved in further implementation of the system. Skepticism and internal resistance, even if directives have covered these in advance, are bad conditions for a successful project.

Simulation of Fine Planning

For operative order planning and sequence optimization, which is a core topic in many projects, simulation can shed light on the matter faster than anything else. Here, it is not sufficient to look at a sample system because every production has different specific conditions and optimization targets. As a basis for simulation, the production structure with the most important parameters and the work plan for a complex article from the product spectrum must be mapped in the MES. The simulation run itself provides information about the timing that can be expected. The result can be compared with actually existing planning and can be assessed accordingly.

Technical Considerations

The degree of sophistication of the functions and the topicality of the technology used are, logically, conflicting requirements. A product in version 1.0 based on the latest technology cannot be "sophisticated" at the same time. In cases of doubt, fulfilling the required functions must be considered primarily.

The flexibility of the system for connection to existing IT systems and especially the option to customize the system are technical evaluation points that also can have financial consequences. A sophisticated release-management function on the system provider side that can guarantee that the system can be updated for many years to come is essential. The investment is sustainable only when the system also can be run for a longer time—the target should be longer than 10 years.

Another technical condition with a financial impact is the licensing model. With a licensing mode that is based on the number of system users, an estimate for the coming years must be made. The *total cost of ownership* (TCO) consideration of the system can be changed considerably by this factor.

Commercial Considerations

The costs of the system can best be compared through an evaluation of the TCO based on the planned running time of the system. In addition to the costs for hardware, software, and services within the scope of the implementation included in the provider's tender, the following costs also should be included in the calculations:

- For planned extensions of use after the actual implementation (e.g., connecting other sections, use of additional modules, additional terminals, additional users for Web-based solutions, etc.)
- For extension and provision of infrastructure (e.g., server, network, etc.)
- For integration with the existing infrastructure (e.g., connection to the ERP system, connection to the LDAP server, etc.)
- For additional training measures
- For regular internal maintenance measures (e.g., database administrator involvement)
- For regular external maintenance measures (e.g., software maintenance contract with update service)
- For user support and hotline

As well as this consideration of the direct costs, the existing deadline planning for the implementation also must be evaluated. Implementation gives rise to indirect costs through the use of internal resources and test runs in production that increase if the time period required for implementation increases. Therefore, a well-thought-out project plan with realistic deadlines and the ability on the part of the supplier to provide the necessary resources are indirect cost factors.

Summary and Decision
After a written summary of the preceding criteria, the decision can be made within the core team. If no clear result arises on the basis of the technical and financial factors, the "soft arguments" should be examined once more. The consultation competence of the provider for strategic instruments such as lean manufacturing and 6Sigma should be given special attention. If the supplier is also competent enough to act as a consultant for these strategic topics, the overall success of the MES can be improved significantly. The company is not just introducing a software system but is "slimming down" or improving processes and thus increasing the overall success of the company.

9.7 Implementation Process

9.7.1 Project Management

Project Management
In addition to the already existing project manager on the customer side, the system provider also must name a project manager and deputy. Both project managers must be released from other tasks in order to be able to dedicate their time completely to implementing the system. These two

key persons also must have sufficient competence to be able to make technical and financial decisions themselves. This means that the risk of delays on the project process is lower.

Project Plan
The project plan defined in the contract specification is verified once more and made more detailed, if necessary. The implementation milestones until approval of the system are established here. All important tasks are allocated to employees at the system provider or within the company. Two key points in the course of a project that are often given low priority (and thus too little time) are the specifications and employee qualifications. A guide for the time required for creating the specifications, especially for project-specific scope and interfaces, is the time scheduled for implementation. The time for specifications thus should be as long as the time allotted for implementation. Only this guarantees that there is sufficient time for including those involved in discussions and forming opinions. The release of production employees for training is difficult during ongoing production. Therefore, these dates and the related organizational regulations must be planned well ahead of time.

Specification
The specification is drawn up by the system provider based on contract specifications and the requirements received. The entire scope not covered by the standard product must be described in particular detail. These details can include project-specific special functions or interfaces with other IT systems, for example. In order to prepare approval, the planned approval criteria should be collected in the form of a numbered list, if possible. This contains the technical requirements and the measurable target values already defined in the contract specification. Before start of the actual implementation, the specification must be released in writing by the customer. The exact establishment and recognition of the scope give the necessary security to both partners. Time-consuming and nerve-wracking discussions about the scope of supply to be expected thus can be avoided.

Regular Meetings
The core team and the project managers should meet for a regular meeting at least once a month, even weekly in the specification and implementation phases. Questions and open points should be clarified immediately, where possible, and if this is not possible, they should be entered on a *list of open points* with the name of the person responsible and the deadline.

9.7.2 Training Management
An MES, like any software system, can be effective only if it is used properly by production employees. This correct use requires, on the

one hand, acceptance of the system (the employee *wants* to work with the system) and, on the other, knowledge about how to use the system (the employee *can* work with the system).

Both wanting to and being able to work with the system are best encouraged through an open information policy and inclusion of employees in all phases of the project, where possible. An equipment operator can, for example, participate in creation of the specification in his or her area and can later support the system provider in the first test runs. System knowledge and acceptance are achieved automatically through this cooperation within the project. However, this cannot replace well-founded, systematic training. Therefore, training plans should be drawn up for all user groups. The documentation should be tailored to the individual groups and also must contain the specific characteristics of the application. "Standard documentation" that does not deal with the special requirements of the application and does not include customizing is not very helpful. Training should be provided for the following user groups, for example:

- *System administrators.* The technical aspects of the system such as installation instructions, data security/traceability, and evaluation of log books should be communicated. A system manual that has already been created is the most suitable basis for this training (see Sec. 9.7.3).
- *Employees responsible for applications.* Employees responsible for applications must be equipped to parameterize MES, extend interfaces, integrate new terminals or new machines/equipment, and create special reports. Thus they must be familiar with all tools of the MES and have a correspondingly high need for training.
- *Production managers and other management personnel.* These persons must be made familiar with all functions that they need for their daily business. These include in particular the planning and information functions of the MES.
- *Maintenance workers.* For this group, alarm management functions, weak-point analysis, and supporting functions for preventative maintenance are of interest.
- *Workers.* In addition to a short introduction to the user philosophy, workers should receive training only on content and functions that are available at their workstations. This training can be repeated at regular intervals but must be provided for all new employees.

9.7.3 Operating Concept

After successful implementation of the system, it must be certain that all functions of the MES are available reliably and that they can be used correctly without the presence of the system provider. An operating concept is necessary that, for example, can be documented in the form of an operating manual. The creation of this concept thus is the last task of the core team and the project manager for this project.

Before the operating manual is compiled, the following questions must be answered:

- *Who is to assume operations of the hardware/infrastructure?* If guidelines are already available within the company for similar systems, these should be applied. Maintenance of the infrastructure can be taken on by internal employees or an external service provider. In both cases, the MES must be entered as a new system in the agreement.
- *Who will install updates and new releases?* Provided that updates and new releases are supplied by the system provider within the scope of a software maintenance contract, there will be some outlay for the installation and testing of the new versions (and even training outlay in some circumstances). Cooperation between internal points and the system provider seems to be the most suitable approach here.
- *Who will assume the first-level application support?* The first point of contact in the event of questions and problems should be an employee within the company (not an external service provider). The task can, for example, be given to the employee responsible for applications. First-level support is also the interface with internal or external support teams and passes on questions that he or she cannot answer himself or herself to these groups.
- *Who will take on second- and third-level application support?* Internal employees theoretically also can cover this service level. However, a higher level of training is required to provide the necessary know-how. Therefore, in practice, these service levels are often covered by the system provider through a support contract.
- *What operating hours must be covered and what support hours (taking into account use in other locations and possible time differences) are needed?*

When these questions have been answered and the user manual has been created, contracts can be concluded with specialized service providers or the system provider, if necessary. Two general types of agreements are usual: a software maintenance contract regulates release management for the software and establishes under what circumstances and in what form new versions of the software are supplied. Installation and testing of the new versions can be regulated optionally within this contract. A support contract regulates answering questions (user support) and solving system malfunctions (hotline support). The required system availability also determines the times at which the supplier can be reached for hotline support. Remuneration for such a contract is made at a fixed-basis amount and an additional variable portion that depends on the actual volumes (number of support cases).

9.8 Summary

At the beginning of this chapter we looked at the implementation of IT systems in general. Selecting suitable hardware generally is easier than selecting software. The advantages and disadvantages of standard and individual software were outlined. Various implementation strategies were introduced in brief, and the problems to be expected as part of implementation were examined in more detail.

Subsequently, the necessary measures (i.e., establishment of a project team, etc.) for preparation of an MES implementation project were introduced. After an analysis of the actual conditions within the company, a detailed project plan was drawn up. Once a suitable system is selected, the actual implementation process begins.

To ensure that the MES implemented is used correctly and accepted as an IT system by all employees, it is especially important to include employees in all phases of the implementation process, where possible, or at least to keep them informed. This is the only way to ensure that the investment outlay in the procurement of an MES is worthwhile. An operating concept secures the investment across the entire running time.

CHAPTER 10
Examples for Application

10.1 Mixed Processes

In the course of research for this book, many examples were examined. A complete integrated manufacturing execution system (MES) as is presented in this book could not be found in any company. Although the need for integrated production management is increasingly recognized, generally, only function-specific isolated solutions for partial areas were found. A clear division according to the production forms

- Discrete manufacturing.
- Process-oriented manufacturing.
- Continuous manufacturing is rather the exception. Generally, mixed processes, that is, process-oriented *and* discrete processes, are needed. The production of roll material, with its refinement process and use-optimized format creation, represents a special form. If we look at the standardized approach for an integrated MES, the borders between the aforementioned production form blur completely. The core topics of the MES always remain the same irrespective of the production form.
- Mapping products in a work plan as a sequence of a combination of work processes and machines/equipment.
- Operative sequence planning across the entire process chain.
- Controlling the production process with
  - Material management (i.e., raw materials, articles produced in-house, and purchased parts).
  - Recording operating data and/or machine data, usually with statistical process control (SPC) functionality.

- Instruction management in various functions.
- Maintenance management for machines and equipment.
- Quality management for machines and laboratory.
- Order traceability.
- Performance monitoring.

The weighting and degree of detail of these requirements differ hugely in individual companies. The examples selected here therefore can only show a small part of the many possible versions. The first example contains a process-oriented mixed process with products being filled into various packing quantities. The second example looks at the characteristics of the production of roll material.

10.2 Sensient Technologies: Emulsions

10.2.1 Information on Sensient Technologies Corporation

Sensient Technologies Corporation is one of the world's leading suppliers of flavors, fragrances, and colors, which are used in a wide variety of products within the food, pharmaceutical, cosmetic, and information technology (IT) sectors. The products are manufactured at various locations worldwide and stand for the highest possible quality. The German location of the company has been certified in accordance with DIN EN ISO 9001:2000 and the International Food Standard.

10.2.2 Description of the Production Process

Emulsion: General Description
An *emulsion* is a semistable solution composed of two liquids that cannot (or only to a limited extent) be mixed with each other. The basic components are water or water-soluble substances and oils or oil-soluble substances. In the emulsion, one of the two liquids is distributed into the other liquid in the form of very fine drops. The dispersed drops of liquid form the *inner phase,* and the liquid that surrounds them forms the *closed* or *outer phase.* These basic substances are also known as *oil phase* and *water phase.*

Since water and oil do not mix by nature, it is necessary to use an *emulsifier* in order to lend the emulsion stability. The emulsifier makes it possible to form a permanent mixture of the water and oil phases by removing the surface tension between both phases.

Production Process for Manufacturing an Emulsion
The water and oil phases are manufactured in-house as preliminary products. There is a similar manufacturing process with a similar method for both preliminary products (Fig. 10.1).

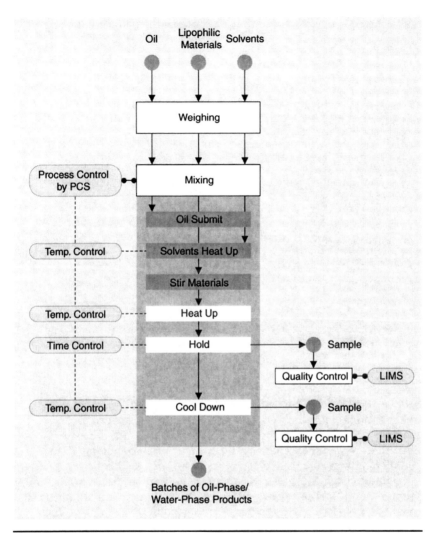

FIGURE 10.1 Process for creating the preliminary product oil phase (the preliminary product water phase is produced in a similar manner).

After the raw materials are weighed, the oil, various lipophilic substances, and solvents are added to a mixing process. This mixing process is carried out and monitored semiautomatically by a process management system (see Sec. 3.4.3) in several stages. At various stages of the process, tests are carried out and checked in a laboratory with the aid of a *laboratory information management system* (LIMS).

Manufacturing these preliminary products occurs in the form of *batches*, which then form the basis for the preparation of the actual emulsion. This production is carried out on special emulsion equipment, which is also controlled by the process control system in the following order:

- Adding the water phase
- Heating the water phase
- Adding the oil phase
- Heating the oil phase
- Mixing the oil phase
- Adding the emulsifier
- Mixing the preliminary products with the emulsifier into a "pre-emulsion"
- Cooling
- Homogenizing
- Pasteurizing
- Cooling
- Filtering (Fig. 10.2)

The finished emulsion then is processed in a bottling system. The actual end product is not achieved until the substance is filled into various container types in the last step; that is, the various articles are determined only through packaging.

10.2.3 Basic Quantity Units and Production Units

The preliminary products water phase and oil phase are produced in the unit of measurement kilograms as a *batch*. Every batch is produced in charges. A *charge* is defined as a *batch lot* and receives a unique ID for control and tracing purposes.

The emulsion is also produced in the unit kilograms in batches with unique IDs. The batches of the emulsion (liters) then are filled into various container forms (e.g., 0.3-liter bottles). Here, then, the unit of measurement changes from liters to pieces. These containers are then packed into boxes, whereby a different number of containers can be used for different packaging types (1 piece to x pieces).

10.2.4 Production Process Plan

The overall production process is mapped in three work plans (for the preliminary products oil phase and water phase and the end product emulsion) by an MES (Fig. 10.3).

10.2.5 Challenges for the MES

Overview
Mixing processes on the basis of recipes are central here. A recipe contains the individual components and their percentages and a process

Examples for Application 201

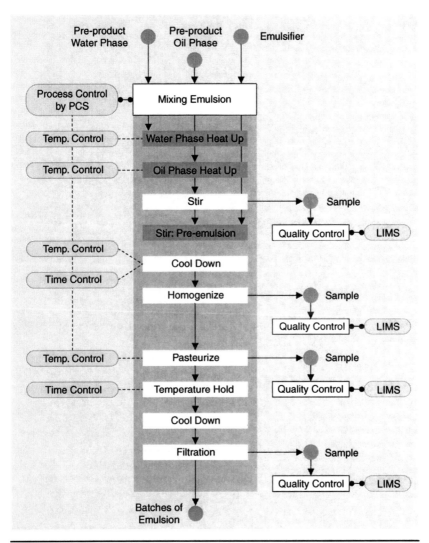

FIGURE 10.2 Process for producing the emulsion on the basis of the preliminary products.

sequence plan, which controls the mixing process. The overall process described is not very complex—the challenge is rather to control the partial systems (e.g., mixer). Here, it is really a partial MES that is used in the sense of ISA S88 (see Sec. 3.2.1). Successive individual raw materials are compounded and added, and the process steps are controlled using a process sequence plan. Because of volume limitations, the quantity is often established and defined in terms of what are known as *charges*. The output arising from a charge then is termed a *batch*.

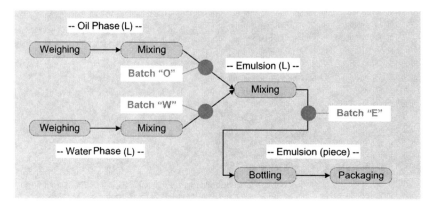

FIGURE 10.3 Macro structure of the production process with production units and quantity units.

In addition to automated control and monitoring of the performance process, samples are taken at certain stages in the process, which then go to the laboratory for testing. If the samples pass the test, the next step is released; otherwise, corrections are carried out.

Special aspects within mixing processes include, for example, waiting periods between individual process stages to allow the product to "mature." Especially for food products, cleaning processes are required that must be carried out precisely according to the Food and Drug Administration (FDA) Directive 21 CFR Part 11 (see Sec. 3.2.4) and confirmed electronically.

These mixing processes generally are preceded by weighing processes in which the individual components as a percentage of the recipe are weighed precisely. For this, regular monitoring of the scales equipment is important to guarantee correct results for the process.

The mixing products generally are filled into a wide range of packing units, which means that a mixture charge is used for several filling charges, differing in packing sizes, labeling, and packaging form.

Although the process steps in process-oriented production generally are highly automated, a superordinate management system is necessary for planning orders and for the order-specific storage, documentation, and evaluation of performance data in a production management system.

In the example described, implementing an MES pursues the following goals:

- Product traceability
- Cycle times—transparent comparison between target and actual times, reduction of cycle times

- Increasing first-pass yield (FPY)
- Increasing capacity
- Increasing overall efficiency (OEE)
- Improving transparency through tracking work-in-process (WIP) stock and the current demand data
- Measurement of deviations from the process sequence plan
- Analysis and evaluation of process parameters (i.e., SPC or 6Sigma)
- Consumption planning and reducing energy consumption
- Detailed resource planning
- Measuring actual production costs online
- Online stock management for warehouse stock, including evaluation
- Maintenance management with assessment of inspection findings and maintenance material
- Increasing planning effectiveness (i.e., planned production versus actual turnover)

The achievement of these goals is closely linked with integration of the process management level, which already exists today in the sense of a collaborative production management system.

Monitoring Function as an Extension of the Automation Level
The mixed results of the individual batches, the associated material application costs, and other performance data with target/actual comparisons with regard to times, energy consumption, and various process parameters must be measured and monitored through the MES. The MES is therefore the superstructure of process automation. The condition for this is a stable and high-performance data interface between the automation and the MES.

Product Traceability
The entire process chain must be documented together with all raw materials used, packing materials, etc. and the relevant process values for the end product produced. By resolving this process chain, starting with the end product, the delivery charges, such as for the raw materials used, can be recorded.

Online Cost Control
A detailed consideration of the cost situation should be carried out for the production of an emulsion, that is, direct production costs, costs for secondary products, warehousing costs, power consumption costs, and overhead.

Performance Recording
For the filling process, an entire overview of the performance situation of the filling system must be provided. This includes target/actual comparisons for quantities, times, and costs, as well as key figures (KPIs) on the efficiency of the system and on the condition of the machines/equipment. In addition, data on the staff employed and on violations of limit values, especially based on statistical quality control (SQC) tests, are also recorded. Individual measurement data can be called up for further analysis.

10.2.6 Realization and Implementation
At the moment, requirements for the MES are being processed as the basis for an evaluation process. The process management systems for the various mixing processes are already in use. A simulation of the process with the aid of an MES gave rise to the following three main requirements:

- Superordinate management and control of the process
- Online cost control
- Performance measurement

These can be mapped with ease. The connection of the existing automation technology is linked with additional outlay.

10.3 Acker: Synthetic Fiber Fabrics

10.3.1 Information on the Company
The company, Acker, was founded in 1949 in Seligenstadt, Germany. From a product line that initially consisted of curtains, technical fabrics were developed beginning in the 1960s. Since the start of the 1970s, these have composed the main branch of production at Acker. Acker is one of the leading manufacturers of technical fabrics.

Acker products generally are supplied to other companies (e.g., automotive suppliers) and then processed further to form products for the end customer (usually not yet the consumer). Examples from the product range are car nets, luggage compartment covers, and fabrics for adhesive plasters and sun screens.

10.3.2 Description of the Production Process

Overview
The production of technical fabrics is carried out entirely in-house at Acker. It generally begins based on an existing need for refinement with the manufacture of what are known as *gray good pieces* (i.e., warping and knitting). Every gray good piece is allocated to an *article*

FIGURE 10.4 Creels with spools of thread as raw material for production.

family and can take on any defined variant (identity = article) within the family in the course of production. The article family establishes the fiber and the construction required for manufacturing the gray good pieces by means of a suitable knitting machine. Production control is handled by sales using what are known as *block orders*. Block orders are, by definition, not article-based (Fig. 10.4).

Refinement collects suitable gray good pieces into *lots*. Sales and marketing assumes production control within the refining process with the aid of *allocation orders*. These are allocated permanently to an article (and therefore have an article identity), but these can be changed throughout the course of production as needed. When a lot is being put together, the kilogram goods are converted to running meters (meterage) using a target key figure. This parameter is based on the periodical recording of actual values.

The batch runs through various production steps within the refining process on the basis of the definition in the *article master*. If necessary, data in the allocation order can be added to the workflow (e.g., the dyeing process step). Each individual step in the workflow is classified using one of the permanently indicated process types (e.g., laundering, napping, and final finishing). The article master data determine with which concrete work cards (i.e., work instructions) the production of a step is to be carried out. The selection of potential production resources is indicated by matching the work card to the concrete production machine.

Accordingly, a work card describes the instructions for the process and the process data for execution of a process step on a particular machine (i.e., equipment). The respective production planning establishes which work card should be used for a concrete production step. Planning is suggested by the MES (i.e., preplanning) and is refined by the production manager (i.e., fine planning or final planning). Manual fine planning is monitored by the MES in order to point out errors or risks. As a result of the planning, production receives the work processes to be executed for the necessary process steps (i.e., sequence of steps).

Depending on the new planning of demand for articles through changes in stock and the content of allocation orders, the article identity of a batch can be adjusted depending on the current production step. Extraordinary planned or omitted production steps to the batch can lead to a change in the article identity.

Warping

If the knitting section orders the warping section to provide chains (i.e., rows of several partial-warp beams), the warping section manufactures these chains and the partial-warp beams needed for the chains on warping machines (Fig. 10.5). Each chain provides several fibers for the knitting process, from which the gray good piece then is manufactured for subsequent refining of the technical fabric through the knitting production step. Eight warping machines are available for warping the partial-warp beams.

FIGURE 10.5 Warping machine.

FIGURE 10.6 Knitting machine.

Knitting

In the knitting section, several partial-warp beams are collected into a chain. A knitting machine is loaded with one or several chains. The production manager assumes planning of the right knitting machine in fine planning. There are 50 knitting machines of different types available (Fig. 10.6). Positioning the fibers into the guide tracks and steering the movement of the guide tracks are done manually. The knitting machine then produces gray good pieces of a particular length and width by means of a meshing technique; these pieces are needed for an article family. The respective gray good pieces are weighed (unit: kilograms) and recorded and called up by refining.

Refinement

The processes of refining involve a lot (a type of batch) composed of several gray good pieces for the required article. The following process steps are (optionally) possible during this process:

- *Pretesting.* The raw material pieces of a lot are subjected to a visual test with the aid of a cloth inspection machine. Flaws in the fabric are marked and recorded statistically. If necessary, various raw material pieces are sewn together.
- *Napping.* In the napping process, the surface of the technical fabric is napped to velour.
- *Laundering.* The gray good piece is cleaned through laundering.
- *Dyeing.* The raw material pieces of a batch are dyed in this dyeing machine.

- *Drying.* A washed/dyed batch is relieved of water residue before the actual refinement process in a stentering frame with the aid of heated air.
- *Finishing.* Finishing refers to treatment of the fiber using a recipe of chemical water-based additives. The finish is applied to the fabric by dipping with subsequent wringing to a defined remaining moisture. In a maximum of five dying fields, the goods are then spread out and condensed out with hot air.

  The finish can be used to achieve various effects, for example, a stiffer or softer feel or flame-retardant effects. It is carried out simultaneously with the process on the stentering frame through a computer-controlled dosing system (Fig. 10.7). The dosing equipment continually provides the stentering frame with the required quantity of finish. The finished goods can be cut optionally in length at the outlet of the stentering frame (symmetric and asymmetric).
- *Fixing.* Fixing is the name given to thermal treatment of the goods with or without finish. It serves in particular to ensure that the goods do not shrink when in use.
- *Final finishing.* The last of one or several work processes is referred to as *final finish.*

  Processes on the stentering frame are controlled, recorded, and monitored fully automatically with the aid of the MES (for quality control). The finished pieces resulting from the

FIGURE 10.7 Dosing equipment.

stentering frame are forwarded to the shipping section for second inspection and tailoring.

- *Shipping.* The workflow for the production of an article ends in refining with final finishing on the stentering frame. Afterwards the lot is sent for shipping. If no second inspection is carried out on the stentering frame, this is performed during a wrapping process on the cloth inspection machine. Flaws are marked and recorded, and the customer is reimbursed for them, if necessary. In the worst case, parts of a lot may be blocked (blocked stock). The finished goods are loaded onto a particular spool type and packed as per the customer's requirements.

- *Laboratory.* In parallel with shipping, a sample piece taken from the lot is forwarded to the laboratory for quality testing. The laboratory checks the characteristics that are specified in the customer order. Inspection criteria such as burning behavior, stretch behavior, and tear strength are also tested in the sample. The actual values measured here are evaluated based on the targets against the tolerances established there (OK/NOK testing). If samples from a lot are tested, shipping waits for the lot to be released before the shipment is released.

10.3.3 Basic Quantity Units and Production Units

See Fig. 10.8 for interrelationships between production units gray good pieces, lots, finished pieces, and spools.

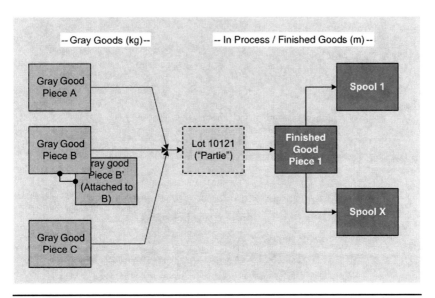

FIGURE 10.8 Interrelationships between production units gray good pieces, lots, finished pieces, and spools.

Chapter Ten

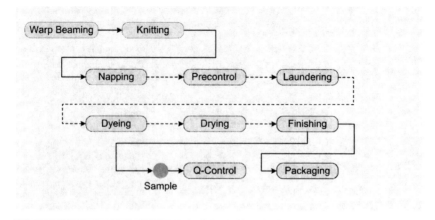

FIGURE 10.9 Macro structure of production flow with production units (i.e., warp beaming, knitting, napping, laundering, and dyeing).

10.3.4 Production Flow Plan

The entire production flow is mapped by the MES in two work plans (for gray good production and refining) (Fig. 10.9).

10.3.5 Tasks of the MES

In addition to increasing transparency and reproducibility of processes and safeguarding processes, the following tasks are also covered by MES.

Master Data Administration

- Article administration with
  - Finishes
  - Work cards for all production departments

Order Monitoring and Confirmation Data

- Order monitoring:
  - Confirmation of operations on production steps
  - Recording of duration of operations
- Resource monitoring:
  - Monitoring of the consumption of raw materials and additives (dosing system)
  - Monitoring of the usage of resources (e.g., spools, dyeing beams)

Operational Data Acquisition (ODA/MDA)

- Linking knitting machines for recording knitting defects (fiber tears) and for the monitoring revolutions
- Linking stentering frames to production management, capture and recording of process values, and usage of finishes
- Linking dosing equipment with usage monitoring of additives and supply of stentering frame
- Linking the drainage system to process recording in order to prove compliance with regulations (for environmental authorities)

Quality Assurance

- Fiber monitoring on warping machine
- Identification of defects on warping machine
- Identification of defects on stentering frame
- Identification of defects in second inspection
- Tests in laboratory and evaluating testing criteria for articles using statistical methods and determining actual values based on customer's testing plans for an article
- Evaluation of test results (laboratory)
- Article-related evaluation
- Customer-related evaluation
- Batch-related (lot) evaluation

Fine Planning and Control

- Production planning for all production departments
- Technical process control

Product Tracing

For the end product produced, the entire process chain with all materials used and the relevant processes must be documented. Tracing will be done with the aid of consistent labeling and monitored process control and recording (e.g., partial-warp beams, chains, gray goods, lots, finished pieces, and reels). Then, for example, it is possible to trace back to the yarn delivery batch used through resolving the process chain, beginning with the end product.

Maintenance (TPM)

- Maintenance and monitoring of mechanical units (roller bearings, etc.) based on recurring examinations
- Carrying out maintenance based on maintenance schedules

- Management of a replacement part warehouse
- Management of testing methods

Recording Production Performance
Identification of the fastest, average, and highest cycle times of the article for all production units

Material Management
- Management of stock quantities for
  - Yarn
  - Warp-beam stock
  - Gray good stock
  - Finished goods stock
  - Blocked stock
  - Additives (dosing system)
  - Work piece carrier/means of transport (reels, beams, etc.)

10.3.6 Challenges

The production described is very challenging, especially because of the wide variety and dynamics this requires. The following boundary conditions must be taken into consideration by the MES:

- Up to the final finishing production stage, however, a lot or part of a lot can take on a different article identity (in the form of a variant) at any time. This is achieved through dynamic redirection within the refining process on instruction. This means, for example, that a lot can be planned under article identity A and take on article identities B and C in part or as a whole in the end result (e.g., through modification of the finish or color).

- The production has both the character of discrete-parts manufacturing (i.e., production of the gray good pieces at the knitting section) and batch manufacturing (i.e., lot), from which packaged goods (finished pieces or reels) result.

- The production may require finishes that are acquired in a continuous process (dosing system) for the manufacture of particular articles. From the point of view of the stentering frame, supply is continual. The dosing system itself, however, can produce the finish only in defined individual quantities (charges). Supply is automatic, and use of the finish with regard to the article must be monitored and recorded (in a calculatory and quality-relevant manner).

- From the preceding, it follows that an existing workflow for an article can be transferred dynamically to the workflow of

another article in accordance with particular criteria. The MES must suggest a transfer where necessary and also must monitor the correct transfer.
- Production is in the form of classic line production. Individual sequences in the workflow can, however, be interpreted as line production (e.g., sequences on the stentering frame).
- Theoretically, the MES must include two partial areas—on the one hand, the knitting section, which provides gray good pieces to refining at the proper time and, on the other, the refining section, which together with the shipping section is responsible for timely delivery of the goods. Within these units, independent coordination is carried out on the basis of internal control.

10.3.7 Realization and Implementation

For realization, a step-by-step process is selected. In the first step, already existing isolated solutions [e.g., production planning and scheduling (PPS) partial systems for scheduling the stentering frame] are extended to integration in the MES interconnections. This migration was urgently needed to retain the existing data records, even if ultimately realization involved replacing some isolated solutions completely with the MES. The tasks of the laboratory are realized in its own software module with its own database owing to its very special nature. This module can be integrated completely into the final MES via a common database.

In the second step, uniform software front ends (based on the .NET framework 2.0) were created for all task areas (i.e., departments) that offer professional customer-specific business processes with the aid of their own databases.

Together with the second step, a standard product for the core of the complete MES was introduced. This standard product assumes, among other things, the required information management, data acquisition (i.e., ODA or MDA), and process linking to controls. Tasks of the standard product also include tracking and tracing, total productive maintenance (TPM) tasks, and resource management. The individual MES terminals use the Web technology of the standard product to implement the required information management. Thus all information and queries are available via Web visualization in the front end on every computer.

In further stages in the process, it is planned to integrate the individual business processes analyzed in advance into the MES system. The number of databases then is reduced from five to two (the MES standard database and a customer-oriented database).

Through the mixture of customer-oriented software development and standard software used, Acker has an optimally adjusted, future-oriented system for its operations. The maintenance costs of the system are manageable owing to the high degree of standard components.

10.4 Summary

In order to illustrate the theoretical content of this book, two application examples were provided from real life. Generally, mixed processes, that is, process-oriented and discrete processes, are required for industrial applications.

The first example includes a process-oriented mixing process that includes filling the products into various packing units. Sensient Technologies Corporation manufactures flavors, fragrances, and dyes that are used in a wide variety of food, pharmaceutical, cosmetic, and IT products. The manufacturing process for an emulsion was described as an example.

The second example examines the characteristics of the production of reel material. The company, Acker, also manufactures products that are used in the production of other products. Examples of some of the company's products are dividing nets for vehicles, luggage compartment covers, textiles for wound strips, and sun screens for cars.

CHAPTER 11
Visions

11.1 Merging the Systems

The production systems in the market today [enterprise resource planning (ERP) systems, product lifecycle management (PLM) systems, and manufacturing execution systems (MES)] have originated and grown from their individual areas of application. A large number of further functions, which are partly covered by other systems on the market, have been added to the originally specified core functions. This means that the systems overlap with regard to their characteristics.

In addition, the systems have different user philosophies, interfaces, and databases that can make them difficult to use. From the user's point of view, it would be desirable to have one exclusive front end available for all functions. The various isolated solutions with their individual data structures and own languages are increasingly merging. This occurs via architectures such as the service-oriented architecture (SOA; see Sec. 7.1.6) or also by means of new products that cover all required functions in one system and make them available to the user. However, it is important to be aware that overly complicated software architectures can only be maintained at great expense.

Another problem is the redundancy of data. This gives rise to increased running and maintenance costs. In addition, where structures are mapped several times, the danger of inconsistent data arises. A consistent data model, in which the product data and the related processes are described in detail and without redundancy, will be essential for future systems. It remains to be seen whether the *master data management* (MDM) method, where different databases can be linked, will succeed in this.

This data model should focus on viable guidelines. Administration takes on a neutral position within the company. All parties involved from all levels participate in this data model (Fig. 11.1). The objects receive a generally valid header data structure to which a large number of detail files for the individual function components (through which individual characteristics can also be mapped) are linked.

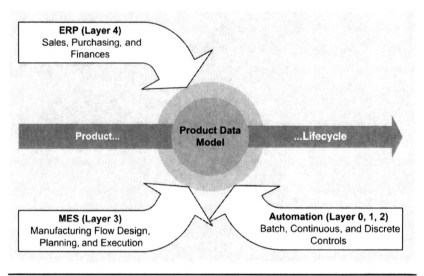

FIGURE 11.1 Central data model for all parties involved in the company.

The header data have the following structure, valid for all objects (products and resources such as machines, materials, personnel, means of transport, operating resources, instructions): identification of the object, phase of the object, description, object group, unit, location, area, date created, and status.

The central object is the product. Unlike other objects, an attribute is allocated to the product that indicates at which phase of its lifecycle (here, we mean the overall cycle, from product concept to recycling/scrapping) the product is.

The product data model therefore must be structured in a way that ensures that it can record all data of the product. It is used to map the entire lifecycle (see Sec. 3.5). This affects the query phase, the conception phase, the phase of actual development, establishment of the process sequence, and finally, the phase of actual production. In the production phase, the product goes through various modification versions until its discontinuation. The lifecycle data themselves, however, remain available in the data archive for further use.

This way of thinking shows how much the individual concepts and worlds merge into each other with this approach. The goal of such an approach must be to eliminate the dominant position of individuals and position the entire company information technology (IT) system on a neutral data framework. This neutral hub makes the required data available to all employees in a well-thought-out structure (see Fig. 11.1).

The origin of the data should be indicated subsequently in an abstract form in the product lifecycle. The product data content grows continually throughout the lifecycle. We distinguish between data from product development and the production process.

11.2 The MES as a Medium of Product-Development Management

11.2.1 Phases of Product Development

The entire product development is a creative value-creation process that is realized by means of a controlled workflow. The trigger for development can have several sources:

- *Technological developments.* Technological developments can lead to new products in companies.
- *Individual customer demand.* In the case of customer demands for new products, it is necessary to check first if the demands can be realized.
- *Market trends.* The market often forces companies to follow new-product trends.
- *Individual ideas.* Ideas that arise internally within the company should be checked to see how feasible they are.

A development phase can be seen as a special work plan that documents the intellectual value-creation process by means of a *development MES*. One phase consists of activities that are carried out with the aid of resources. These can be a single person, a team, or development tools. Schedule times for the activities are allocated to them. The development process of the phase itself consists of intellectual value-creation activities such as deliberation, information searches, discussion, design, and finally, documentation.

Recording of the creative value-creation process is carried out according to the principles of an MES. For the development order, the performance data (i.e., resource usage and time usage) per phase is recorded, monitored by permanent target/actual comparison, and then documented. The individual phases can be seen in Fig. 11.2.

11.2.2 Request Handling

If a potential product development is based on a customer request with individual requirements that should result in a quote, the work plan consists of the work steps *handling the order request* and *creating a quote*. For this "new" potential product, the header data set of the object *product*, valid for the lifecycle, is assigned in this phase. This data definition activates sales and marketing and compares it with the neutral position that is responsible for the product data model.

Since creative value creation is carried out mainly in the development process, this process must be structured and documented to be

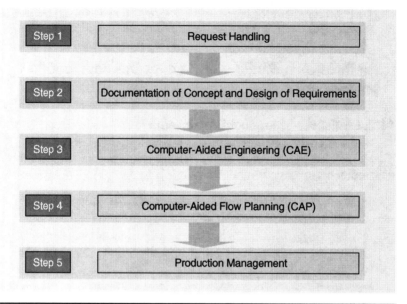

FIGURE 11.2 Product-development phases.

reusable. Here, the requests from the customer side are first recorded in a standardized document. Documentation and quotes are filtered out of the product data system via an intelligent filter system in order to make it possible to create a quote quickly. In the second work step, the actual calculation follows, in which a quote is created by means of the requirements recorded and a generally valid calculation system. If an order ensues from a quote, the actual development is initiated. If the sources for the development of a new product are a new technology, the market (i.e., needs or necessities), or an individual idea, the development process begins with phase 2.

11.2.3 Concept Documentation and Designing Requirements

This phase can be divided into two work steps, which are reflected in the work plan. As with an actual production process sequence, processing planning times are allocated to each step. The individual workplaces are equivalent to *team centers*, for example.

The first step processes the concepts and requirements. Processing materials are used as material usage or the documents created in phase 1 are transferred if an order has been placed based on a quote.

Within the processing of concepts, the individual concept contributions of members of the team center are collected, structured, and as described earlier, saved in standardized documents. When this work step is completed, the result of the first step is transferred to a requirement concept in the next work step, *design of requirements,* which compiles all specifications for construction by means of three-dimensional (3D) computer-aided design (CAD) and digital mock-up (DMU) systems.

A significant step for this is filtering out information from the product database. Here, the focus is on finding already manufactured articles with properties similar to those of the product to be developed.

The search criteria are often an article group in which a particular material is used, where a particular operation/machine combination is required for the articles, or where articles have similar geometric data. This whole process should be extended to a range of other characteristics as search criteria. This means that a great deal of time can be saved because it is possible to resort to components that have already been processed.

The result of the conception phase is compiled once more in standardized documents and saved in the product database. Before this phase is complete, the team should check the documents once more; then this phase is released, and the status of the article is changed.

11.2.4 Construction of the Product

In this phase, the product that is to be newly developed is developed into a virtual product (DMU = digital mock-up) using 3D CAD and finite-element calculation on the basis of the information provided. These sophisticated tools play a major role, especially in the assembly processes and their simulation and are supplied by special providers and integrated accordingly. With DMU, the development process is accelerated considerably because, for example, real prototypes (as opposed to virtual prototypes) generally can be avoided.

DMU provides a wide range of data for the development of real process sequences such as data for numerical control, part list data, quality assurance (QA) data, and production instructions. In chemical processes, data transfer occurs without CAD tools in test laboratories. Here, recipes and their sequence procedures are developed, adhering to the permitted substance content. All these data are transferred to the next phase of production planning.

The work plan for the CAD phase is divided into the work processes for the actual construction and the consolidation of the data for the subsequent process of process sequence planning. The work steps must be freely definable. Every work step is allocated a schedule time for processing. The documents of the predecessor are allocated as *material usage,* as operating resources for the corresponding CAD program, whose time usage is evaluated from the cost aspect and flows into cost-control management. The processor can choose from the instructions that have been handed over by the predecessor in the conception phase. These are displayed to the processor.

When the CAD program is selected, the system is started, and the construction phase begins. The results and the data on the geometry of the product generally are managed in an object relational 3D database with a corresponding index structure. This supports multileveled request handling and thus contributes significantly to the acceleration of geometric requests for DMU and to the search for similarity.

At the end of the work step, instructions are created as in the previous phases. For the individual functions, the corresponding documents are selected from a list of standardized formulas in order to enter the data that are important for the successor.

11.2.5 Computer-Aided Flow Planning

Attempts have been made time and again to draft a work plan automatically from the DMU data. The results to date have been modest because too much data have had to be added manually. This includes data for process flow sequence, setting of time targets, employee qualifications, material selection and allocation, facilities, quality, etc.

Work planning consists of the work steps *development of the process flow with its resource usage, virtual simulation,* and if necessary, *actual prototyping*.

In this phase, the required flow procedures are developed in machines and systems [e.g., computer-aided manufacturing (CAM)]. These are numerical control programs for discrete processes and flow procedures of recipes in batch-oriented processes. The interaction of the individual phases also can occur retroactively. For example, if the worker changes the numerical control program in the production phase for tolerance reasons, it must be decided whether this was necessary only for the current production charge or whether there is an error that must be changed in construction. However, the numerical control data are derived from the CAM model. This, in turn, is based on a CAD diagram.

The results are documented separately in initial sample-checking reports and are handed over to the customer with the samples. The entire process takes place within the framework of the *product part approval process method*. After release by the customer, an extensive product description then exists for the actual production process.

It is apparent from the phase concept that the product description is refined continually via the phases until the approved process flow description for the product can be assumed in real operations. The individual contents of the product data model were described in Chap. 4.

Regardless of what concept is used for this merged IT system, the development management contains significant aspects of an MES, albeit in a somewhat modified form. It supplies the following significant results:

- Production flow-oriented design as the basis for a production MES.
- Comprehensive performance recording and monitoring of product development.

- Recording of development costs for the product, which flows into the new calculation systems with product-based distribution of the overhead costs.

11.2.6 Production Management

When the development data from the work plan are transferred into the actual MES, the foundations are laid for the planning and recording of the actual performance process. The product master data for the production process are maintained in this phase. Changes are made, which are maintained in different versions of the work plan. These can be modifications that do not affect the development side, for example, changing tolerance or action limits, as well as modifications that are important for the development side because they carry out changes in the process flow that must be coordinated with the construction section. These data are documented in a production history.

It is worth mentioning that finance management (today, the dominant position in company IT) becomes active in the lifecycle of a product only when orders are generated and thereby significant commercial data such as prices, goods accounts, and cost data from the operational cost accounting section, etc. The lifecycle of a product ends when it is discontinued, for example, because the market no longer has any use for it or because it is replaced by a more modern, more cost-efficient product.

11.3 Standardization of Function Modules

Through a consistent data model, it is also possible to standardize *function modules* (i.e., part list function, material requirement planning, personnel deployment planning, etc.). This contributes to a rationalization of the flows.

One example is the planning that takes place under different aspects. One aspect is strategic planning with the core focus on the consideration of middle- and long-term resource requirements. Today, this function is offered by all ERP systems. Another aspect is operative planning with the focus on sequence optimization with regard to the short-term resource requirements. This function is possible today only with an integrated MES, as described in Chap. 5.

In the future, there will be one planning system exclusively that will do justice to both aspects; thus data and information alignment becomes unnecessary when planning is changed.

11.4 Merging Consultancy Activities and IT Systems

Integrated systems of this kind require appropriate quality on the part of the user, the consultant, and the software provider. Companies that deal with the requirements of the factory of the future and plan strategic rationalization initiatives will invest increasingly in employee training.

The consultancy side will expand its services in order to be able to provide qualitative advice regarding MES ("Lean Sigma"). Software providers must face the challenges of integrated systems through new developments. Only the combination of analysis/evaluation and sophisticated products will yield optimal results.

11.5 Summary

We have tried to sketch an integrated product data structure that is neutral and is available to all users in the company to the same extent. This means that the systems merge with regard to their functionality and use a common database. For this, it is also essential that the structure of the data, which have been mapped in several tools thus far, is standardized.

The functional separation of ERP system and MES was clearly possible in the previous chapters. However, there is also the issue of the common database, which today is usually mapped redundantly. Approaches such as SOA and MDM will defuse this topic.

The visions presented contain no statements as to which existing systems probably will be replaced by others. The intent was only to show the need for action to achieve harmonization and provide an approach for a solution.

CHAPTER 12
Summary of the Book

In order to be able to compare the systems on the market and assess them with regard to their suitability, the relevant basic knowledge about manufacturing execution systems (MES) has been communicated from a neutral standpoint. Decision makers in a company thus have the necessary means in their hands to find a suitable system and be able to assess the investment during procurement with regard to its profitability even before implementation.

This book is suitable as an introduction to production-relevant information technology (IT) systems owing to both its neutral presentation and its extensive handling of the topic of MES. Extended information about technologies, implementation potential, etc., was also imparted so that it is not necessary to rely on previous knowledge. Below is a brief summary of the content.

The MES is developing into a strategic tool for flexible and networked production. All production management tasks are collected in an integrated platform. As a database, an MES requires a thorough and consistent data model that contains the product data as well as a memory map of the production with all resources. Thus the MES must be closely linked to the product lifecycle management (PLM) system and work hand in hand with it. In order to be able to fulfill the tasks of the factory of the future, additional new functions are needed. This makes the MES a central strategic tool.

Before an approach is developed for the requirements identified for the factory of the future in the core chapters, existing standards and technologies that can be used to solve the problem should be considered in terms of quality. Analysis of the relevant norms and directives shows that there are some approaches to the subject area of the MES. However, on closer examination, it becomes apparent that all norms, directives, recommendations, etc., are based to a great extent on ISA S88 and S95.

An MES as a tool for the production management level requires a complete product definition for its task. The overall administration of the product data should be assigned to the MES. The product definition data from the MES must be available to all other applications through transparent structures and suitable software technologies.

At the core of the data model is the work plan, which describes the production process of an article as a sequence of activities/processes with all required resources. Since large technological and structural differences exist between different products, a universal data model must be constructed to be flexible and expandable. In particular, aspects of a *mixed production* must be provided with process and discrete production steps, and the different *quantity units* associated with them must be considered.

Furthermore, a qualified MES has an operative planning system that plans a collision-free order pool, taking into consideration the resource situation. The system ensures that order processing is subject to a realistic framework and that possible deviations are recognized immediately through a constant target/actual-value comparison. Fundamentally, such a system must contain planning algorithms for optimization that are based on realistic work plan data. The result of planning calculations should be displayed graphically in a Gantt diagram for a better overview.

In order processing, various sections of the company are involved, both directly and indirectly. An MES gives everyone in these sections a suitable tool that makes it possible to carry out the work in the best possible way and guides the employee with the aid of a defined workflow.

In production, the orders to be processed are displayed to the machine operator via terminals. All required additional information is visible online. Furthermore, all relevant data from the machine are passed on to the MES for evaluation via appropriate ports. Other production-related areas such as logistics and maintenance also source the necessary information and work orders from the MES. The prepared data are provided, in turn, to the production manager, the controlling section, and business management in a suitable format. Data access occurs by means of Web technologies throughout the company.

After introduction of the core functions of an MES, we examined the technical possibilities and their application in more detail. The general software architecture of an MES, central components of the system, etc., were explained. The core of an MES is formed by a powerful database. The conditions for this and the necessary technologies were described. Here, measures for archiving play just as important a role as the ongoing maintenance of relevant systems; only this guarantees optimal use of the MES. Subsequently, there was a detailed consideration of the interfaces, both to other systems and to the user. Various technologies and mechanisms for communication that are used today in the MES environment were considered.

Later in the book, existing methods for intentionally reducing sources of loss in production were explained. These methods are mostly possible only with the support of a suitable IT system. As shown in the explanations, an integrated MES can be of great use for a production company with regard to reduction of the sources of loss shown. Professional project implementation is essential to ensure the success of the project.

Selection of the appropriate hardware generally is simpler than the choice of software. The advantages and disadvantages of standard and individual software were outlined. Various strategies were explained in brief, and the problems that can be expected in implementation were examined more closely. Necessary measures (e.g., forming a project team, etc.) for the preparation of an implementation project for an MES were presented.

In order to ensure that the MES implemented is used correctly and also is accepted as an IT system by all employees, it is especially important to involve employees in all possible phases of the implementation process from an early stage or at least to keep them informed of same. Only this guarantees that the investment made in procuring an MES will pay off.

In order to be able to demonstrate the theoretically communicated content of an MES to the reader in a clear way, two examples of applications in practice were provided. The first example consisted of a process-oriented mixed process involving filling the products into various containers. The second example examined the characteristics of the production of real goods.

To close, visions of the future form of production-related IT systems were described. Among other things, an attempt was made to sketch a universal data structure that is neutral and is available to all users within the company to the same extent. It was predicted that the existing systems will merge.

References

[ANDERL 2002] Anderl, R. *Produktdatentechnologie 1. Hypermediales Skriptum zur Vorlesung.* Internet source URL: www.iim.maschinenbau.tudarmstadt.de/pdt1/frames/PDT1.html.

[ARC 2003] ARC Advisory Group. *ARC Reference Sheet—Collaborative Production Management.* Internet source URL: www.arcweb.com/Brochures/Collaborative%20Production%20Management%20Ref%20Sheet.pdf.

[ARNOLD ET AL. 2005] Arnold, V., et al. *Product Lifecycle Management beherrschen—Ein Anwenderhandbuch für den Mittelstand,* 1st ed. Springer-Verlag, Berlin, Heidelberg, 2005.

[BALZERT 1998] Balzert, H. *Lehrbuch der Softwaretechnik—Software-Management, Software-Qualitätssicherung, Unternehmensmodellierung.* Spektrum Akademischer Verlag, Heidelberg, Berlin, 1998.

[DIN 19222] DIN 19222: *Leittechnik—Begriffe.* Beuth-Verlag, Berlin, 2001.

[DIN 19233] DIN 19233: *Prozessautomatisierung—Begriffe.* Beuth-Verlag, Berlin, 1998.

[DIN 44300] DIN 44300: *Informationsverarbeitung—Begriffe.* Beuth-Verlag, Berlin, 2000.

[ELPELT HARTUNG 2007] Elpelt, B.; Hartung, J. *Multivariate Statistik: Lehr- und Handbuch der angewandten Statistik,* 7th ed. Oldenbourg Wissenschaftsverlag GmbH, Munich, 2007.

[FRÜH MAIER 2004] Früh, K. F.; Maier, U. *Handbuch der Prozessautomatisierung—Prozessleittechnik für verfahrenstechnische Anlagen,* 3d ed. Oldenbourg Industrieverlag, Munich, 2004.

[IEC 61512] IEC 61512: *Chargenorientierte Fahrweise—Teil 1: Modelle und Terminologie.* Beuth-Verlag, Berlin, 1999.

[IEC 62264] IEC 62264: *Enterprise-Control System Integration, Part 1: Models and Terminology.* Beuth-Verlag, Berlin, 2003.

[ISA S88] ISA S88-1: *Batch Control.* ISA, Research Triangle Park, N.C., 1995.

[ISA S95-1] ISA S95-1: *Enterprise-Control System Integration, Part 1: Models and Terminology.* ISA, Research Triangle Park, N.C., 2000.

[ISO 9000] DIN EN ISO 9000: *Quality Management Systems—Fundamentals and Vocabulary.* International Organization for Standardization, Geneva, 2005.

References

[ISO 9001] DIN EN ISO 9001: *Quality Management Systems—Requirements*. International Organization for Standardization, Geneva, 2000.

[ISO 10303] ISO 10303: *Industrial Automation Systems and Integration*. International Organization for Standardization, Geneva, 1994.

[KAPLAN NORTON 1992] Kaplan, R. S.; Norton, D. P. The Balanced Scorecard: Measures That Drive Performance. *Harvard Business Review*, Jan.–Feb. 1992.

[KRUMP 2003] Krump, F. *Diffusion prozessorientierter Kostenrechnung*, 1st ed. Deutscher Universitäts-Verlag/GWV Fachverlage GmbH, Wiesbaden, 2003.

[MCCLELLAN 1997] McClellan, M. *Applying Manufacturing Execution Systems*, 1st ed. CRC Press, Boca Raton, Fla., 1997.

[MESA 1997] MESA International. *The Benefits of MES—A Report from the Field*. Manufacturing Enterprise Systems Association, Chandler, Ariz., 1997.

[MODBUS 2007] Modbus-IDA. *Homepage*. Internet source URL: www.modbus.org.

[NE 59] NE 59: *Funktionen der Betriebsleitebene bei chargenorientierter Produktion*. NAMUR, Leverkusen, 2002.

[OPC 2008] OPC Foundation. *Homepage*. Internet source URL: www.opcfoundation.org.

[PNO 2007] Profibus Nutzerorganisation e. V. *Homepage*. Internet source URL: www.profibus.com.

[SYSKA 2006] Syska, A. *Produktionsmanagement—Das A-Z wichtiger Methoden für die Produktion von heute*, 1st ed. Gabler-Verlag, Wiesbaden, 2006.

[VDA 2008] Verband der Automobilindustrie. *Homepage*. Internet source URL: www.vda.de.

[VDI 2219] VDI 2219: *Informationsverarbeitung in der Produktentwicklung—Einführung und Wirtschaftlichkeit von EDM/PDM-Systemen*. Beuth-Verlag, Berlin, 2002.

[VDI 5600] VDI 5600: *Fertigungsmanagementsysteme*. Beuth-Verlag, Berlin, 2007.

[VDMA 2008] Verband deutscher Maschinen- und Anlagenbauer. *Homepage*. Internet source URL: www.vdma.de.

[W3C 2007] Word Wide Web Consortium. *Homepage*. Internet source URL: www.w3c.org.

[ZARNEKOW BRENNER PILGRAM 2005] Zarnekow, R.; Brenner, W.; Pilgram, U. *Integriertes Informationsmanagement*. Springer-Verlag, Berlin, 2005.

[ZIMMERMANN 2005] Zimmermann, Z. *Möglichkeiten zur Unternehmensweiten Harmonisierung von Stammdaten*. Grin-Verlag für akademische Texte, Munich, 2005.

Glossary

ActiveX Microsoft technology used to integrate external applications as components into your own program.

Activity-based costing (ABC) Activity-based costing is a process in which the average costs of production are allocated in an article-specific manner. In this approach, it is assumed that the company's resources are used for providing services (for production in the sense of the MES). The costs of the resources are allocated to the activities that require these resources. The sum of the resource costs of an activity forms the activity costs. The costs of the product are derived from the sum of the activity costs. Here, we refer to all cost elements that cannot be allocated as direct expenses in particular to the production overhead costs. Thus, where activity-based costing is in use (as opposed to various other costing approaches), production is expressly included.

ADO Abbreviation for "ActiveX data objects."

AJAX Abbreviation for "asynchronous JavaScript and XML." Concept for the creation of Web applications combining the advantages of a lean Web solution (no applets or other objects that need to be loaded) with the agility of a rich client. To put it simply, only modified data are exchanged between the client and the server, and only modified elements of the views used are updated.

Andon board The concept *andon* originated in Japan and is a system for triggering improvement measures. With the aid of andon systems, workers can release optical or acoustic signals, for example, in the case of quality problems or malfunctions. In today's usage, an andon board is a display system that is usually located under the ceiling of the hall and displays status information from production area across a wide distance.

ANSI Abbreviation for "American National Standards Institute."

API Abbreviation for "application programming interface." Documented interface for the integration of functions based on library elements in own applications. For example, the Win32-API allows the use of Windows functions.

APS   Abbreviation for "advanced planning and scheduling." Extends the planning strategies of an ERP system so that several orders can be planned without collision in the operative field of production, taking into account the availability of resources. With the planning algorithms provided in APS, an optimal sequence for processing the orders according to priorities and rules is also provided (see also *Fine planning*).

ASCII   Abbreviation for "American Standard Code for Information Interchange." Character set that contains just 128 characters, the first 33 of which are functional characters (e.g., line break).

ATP   Abbreviation for "available to promise." Goods are available to promise if a binding delivery date can be confirmed to the purchaser. This can be ensured through APS (advanced planning and scheduling).

Audit   Investigative processes for evaluating products or process sequences with regard to the fulfillment of requirements and directives are generally referred to as *audits* (from the Latin for "hearing"). Audits are carried out by an auditor who has been specially trained for the purpose. For example, the TÜV (Technischer Überwachungsverein or German Technical Inspection Agency) carries out audits for DIN EN ISO 9001:2000.

Availability   Availability as a percentage indicates for what period of time related to a defined overall time (planning bin occupancy period) for which a machine is actually ready for production, that is, available. Sharing in the total loss (not ready for production), for example, are down times owing to malfunctions or setup times.

Balanced scorecard   The *balanced scorecard* is a concept for the balanced and implementation-oriented control of production with the aid of parameters. The performance of an organization is seen here as a balance between the financial management, the customer, the business processes, and employee development and is represented in a clear form (scorecard).

Base quantity unit   This quantity refers to the article to be produced and is used as a reference quantity for all descriptions and calculations in the work processes. For example, the time specifications, number of preliminary products used, or required quantity of raw material always refers to this base quantity unit. In discrete production, the base quantity unit is one piece. In procedural processes, this unit can be 10 kg or 100 L, for example.

Batch   Batch is a frequently used term for production or delivery units. Similarly to orders, batches are identified using unique IDs. Generally, a production or procurement time and a quantity are associated with them. Often a batch is a subunit of an order. Order tracing also can be carried out using a batch number.

Benchmark   Comparable value or target value for measurement and orientation in a ranking list. For example, a company compares its profit margin with the market leader, which sets the benchmark in this case.

Bill of material English term for parts list.

Bill of process English term for work plan.

CAD Abbreviation for "computer-aided design."

CAE Abbreviation for "computer-aided engineering." The term CAE summarizes all possibilities of computer support in development. The concept can be understood similarly to CAD, which is part of CAE. As well as modeling and conception, CAE also includes advanced analyses, simulations of many physical processes, and optimization tools.

CAM Abbreviation for "computer-aided manufacturing." CAM is the support of production through automatic processes that run in the machine/equipment such as NC programs or sequence procedures in batch processes.

CAP Abbreviation for "computer-aided process planning." This planning builds on conventional or CAD-based construction data to create work plan data.

Capacity Quantities or services that can be achieved in a defined period.

Capacity accounts A capacity account is managed per resource. On the capacity account, inflows and outflows of capacity are booked. The balance of the capacity account thus provides current availability and capacity of the resource.

Capacity planning Planning of resource usage based on the current capacity account status and the orders planned.

CAQ Abbreviation for "computer-aided quality assurance." See also *Quality management, SPC, SQC,* and *TQM*.

CI Abbreviation for "corporate identity." Refers to the entire public image of a company.

CIM Abbreviation for "computer-integrated manufacturing." CIM denotes the integrated EDP (electronic data processing) application in all divisions connected with production; CIM is composed of the components CAD, CAE, CAM, CAP, and CAQ.

COM/DCOM Abbreviation for "Component Object Model/Distributed Component Object Model." Technology developed by Microsoft for communication between objects/processes on a Windows platform. DCOM extends COM with networking options, and this makes networked use of the COM objects possible in a distributed system.

Cp/Cpk Cp and Cpk are parameters that express whether a (production) process is stable or controlled. Cp stands for "process capability" (and Cpk for "process capability index"). If the quality characteristics are distributed only randomly, and if the development (process situation) is solely within the action or tolerance limits, the process is referred to as *capable*. These parameters often replace the machine capability index because too many influencing variables affect machine capability (e.g., means of measurement, environment, employees, etc.). A process is deemed capable with regard to a characteristic if this characteristic of a production part lies within given tolerance limits with a probability of 99.63 percent (±3x standard deviations).

CPM   Abbreviation for "collaborative production management." In a CPM system, the functions of an MES are combined with other components such as ERP, CRM, and SRM; that is, the individual function components work together. Cooperation in this regard also can occur between programs of different manufacturers.

CRM   Abbreviation for "customer relationship management." CRM systems are software systems for managing the relationships between suppliers and customers. All data of these relationships, beginning with requests and order processing, are recorded. Specific evaluations of these data form the basis for marketing and for optimizing the range of services offered. A CRM system is located in the company management level as an independent system or is an integrated component of an ERP system.

CSV   Abbreviation for "comma/character-separated values." List of values separated by a unique symbol.

Cycle time   The period between the beginning of the first activity and the end of the last activity based on a defined activity sequence. Within production, cycle time denotes the time period needed from the beginning of processing (start of a work process in the work plan) to the completion of a product (end of last work process in the work plan). The cycle time is composed of setup time, processing time, and waiting time.

Data warehouse   A data warehouse is a central database, usually in the form of a relational database, the content of which is composed of data from various sources. The data are loaded to the data warehouse from the data sources and made available to all applications in the company centrally from here. The data are saved and are used mainly for data analysis and as an aid for management decisions.

Days of inventory   Quotient from the current stock and the average usage of a resource or article per day. The measured quantity indicates how long the stock will last (in days).

DBMS   Abbreviation for "database management system."

Degree of utilization   The degree of utilization is defined as the ratio of the productive running time to the possible utilization time of the machine within a defined period of time. The degree of utilization is similar to the performance rate.

Glossary

Digital mock-up (DMU) Digital mock-up is a continuation of the CAE concept. As early as possible in the construction phase, it provides a complete virtual digital 3D model of the product. In digital prototype construction, the primary focus is on virtual assembly and on construction analyses from simple component groups to complex product structures. Analyses on the basis of 3D CAD models include especially checking for freedom from collision, adhering to minimum distances, suitability for assembly, etc.

Dispatching The technical term dispatching in the context of an MES refers to sequence construction and distribution of tasks and functions to machines/equipment.

Disposition Disposition within the context of an MES refers to the selection, reservation, and planning of materials and resources for production. For example, material and workers are allocated to a planned order.

DMAIC DMAIC stands for "define, measure, analyze, improve, and control" and defines a closed circuit for implementing measures. Define = recognize and define the problem; measure = measure related parameters; analyze = analyze the results; improve = improve the situation; control = control the measures taken using new measurements. This concept aims for a 6Sigma production.

DNC Abbreviation for "distributed numerical control or dynamic numerical control." In manufacturing technology, describes embedding computer-controlled machine tools (CNC machines) into a computer network. Where necessary, the processing programs (NC programs) are loaded into the control of the machine by one of the connected computers with the aid of the DNC system.

EAN code EAN stands for "international article number" (formerly "European article number") and is a product description for commodities in the form of a barcode. EAN is a number composed of 13 or 8 figures that is managed centrally and is assigned to manufacturers on request. The code contains the following data for the identification of an article: country identification, user number, and article number.

EDI Abbreviation for "electronic data interchange." Electronic data exchange, generally interplant, whereby the definition does not give any information on the standard protocol used (e.g., UN/EDIFACT).

EDM Abbreviation for "engineering data management." EDM includes the holistic, structured, and consistent administration of all processes and data accumulated in the development of new or in the modification of existing products throughout the entire product lifecycle.

EJB Abbreviation for "enterprise JavaBeans." Technology for mapping functions within an application server, originally developed by IBM and SUN based on Java.

Glossary

ERP   The concept enterprise resource planning (ERP) denotes the corporate task of planning the resources within a company (e.g., capital, facilities, or personnel) in the most efficient manner possible for the operational processes. ERP systems should map all business processes. Typical functional areas of ERP software are material management, accounting and finance, controlling, human resources management, research/development, and sales/marketing.

Event management   Events in production are, for example, actions of the user, warning messages/alarms from a machine, and control messages from a software module of the MES (e.g., SPC). Event management takes on the processing and reaction for these events.

FDA   The Food and Drug Administration (FDA) is a public U.S. agency of the Department of Health and Human Services. It is responsible for the safety effectiveness of human and veterinary medicines, biologic products, medical products, foodstuffs, and ray-emitting devices. It supervises the manufacture, import, transport, storage, and sale of these products. Implementation of its guidelines should ensure that the American consumer is protected. This applies for products manufactured in the United States as well as products imported into the United States. Therefore, these guidelines are also binding for European companies that export to the United States.

FDA 21 CFR Part 11   The directive 21 CFR Part 11 of the FDA defines criteria for which the FDA accepts the use of electronic data recording and electronic signatures as equivalent to data recording and signatures on paper. Here, the focus is the quality assurance process in the area of the foodstuffs and pharmaceutical industries.

FIFO principle   FIFO stands for "first in, first out" and describes a principle for administering buffers; the object that was first booked into stock is also the first to be booked out.

Fine planning   Operative order planning is the core function of an MES and is also often referred to as *fine planning*.

FMEA   FMEA stands for "failure mode and effect analysis" and is a method used to recognize potential error sources at an early stage (in the development process) and avoid them through suitable measures. However, it is also used in ongoing processes. We distinguish between construction and process FMEA.

FPY   Abbreviation for "first-pass yield." FPY is a measurement for quality assurance that indicates the share of products not needing reworking as compared with the entire quantity produced.

HRM   Abbreviation for "human resources management." Software function for administration, control, and wage accounting for a company. HRM is a function usually integrated into an ERP system.

HTTP Abbreviation for "Hypertext Transfer Protocol." HTTP is the conventional protocol for the exchange of data via the Internet.

ID Abbreviation for "identifier." An ID serves for the clear identification of an object (e.g., serial number of a device or chassis number of a car) in a class of objects.

Incoming goods inspection Incoming goods inspection is a partial function of the MES and includes the commercial, technical, and material testing of incoming goods.

Instruction management Instruction management in production includes the management and provision of information for workers via display systems or in paper form. In the scope of the MES, these are, for example, installation instructions for tools, process instructions, or machine settings.

Inventory turnover Inventory turnover is a parameter for the effectiveness of production and indicates how often the average value of the stock is turned over per year.

IPC Abbreviation for "industry PC." Robust computer designed for use in extreme environmental conditions (with regard to temperature, vibrations, moisture, etc.).

ISA abbreviation standing for the "Instrumentation, Systems, and Automation Society." The organization is internationally active and currently has more than 28,000 members in more than 100 countries. The tasks and goals of the organization include drawing up guidelines on measurement technology, controlling and regulating processes, and organizing congresses and trade fairs on these topics.

ISA S88 Part 1 of the S88 guideline of the ISA defines reference models for batch controlling of the process industry, relationships and ratios between the models, and processes. In Part 2, data models and their structures for batch controls in the process industry are defined, which should make it easier to standardize communication within and between the individual batch controls.

ISA S95 Part 1 of the international standard ANSI/ISA S95 describes the fundamental terminology and models with which the interfaces between the business processes and the process and production management systems can be defined. Part 2 defines the interface content between the control functions and the company management. Part 3, which appeared in 2005, provides detailed definitions of the main activities of production, maintenance, warehouse maintenance, and quality control sections.

Just-in-sequence (JIS) An inventory strategy in which parts and preliminary products arrive at a production line moments before they are needed and in the order needed for synchronized production (usually an assembly process).

Just-in-time (JIT) A concept for material supply that aims at reducing interim stock and generally streamlining the production process. Parts and preliminary products are provided to the production line with minimal time in advance.

Kaizen *Kaizen* is a Japanese concept (*kai* = "change"; *zen* = "for the better") and consists of measures for continual process flow improvement.

Kanban Japanese concept meaning "accompanying card" and denoting a system for the control of part supply using the pull principle with a goal of lower on-site stocks. The point of consumption reports a need to the point of supply by placing an empty container at a defined handover location indicating the type and quantity of the article needed.

KPI Abbreviation for "key performance indicator." A KPI is a parameter that reflects the degree of fulfillment in regard to targets set or a critical success factor. The best known examples from the MES area are availability, quality rate, and OEE (overall equipment efficiency).

LDAP Abbreviation for "Lightweight Directory Access Protocol." LDAP is an access process to a directory server used for the central administration of user accounts, for example.

Lean manufacturing The key phrase *lean manufacturing* goes back to the measures drawn up by Toyota to increase production efficiency in the 1950s. The concept, known under the term TPS (Toyota production system), mainly involves avoiding and reducing sources of loss, especially waiting, idle, and transport times. An MES is an important tool for implementation in addition to conventional methods.

LIMS This abbreviation stands for "laboratory information management system."

Lot/lot size A lot is a frequently used term for production or delivery units. Similarly to orders, lots are identified using unique IDs. Generally, a production or procurement time and a quantity (→ lot size) are associated with these. Often a lot is a subunit of an order. Order tracing also can be carried out using a lot number.

Make-to-order English concept for contract manufacture. Products are produced only after receipt of an order. This reduces warehousing costs.

Make-to-stock English term for stock production. Production generally occurs in an anonymous market in which an inventory is produced and warehoused based on a demand determined. The warehousing costs are higher than in the make-to-order method.

Material management Material management in production involves management of the materials required for the production process and order-based use. The overall procurement system is integrated. In addition to stock management for the raw material, production warehouses for partial products are also managed.

Maximum stock level   The maximum stock level is the highest level of stock that may be available in a warehouse for capacity-limitation reasons.

MDA   Abbreviation for "machine data acquisition." MDA stands for the online capture of reports, measurement data, and parameters from machine controls through superordinate systems. MDA is a partial function of a qualified MES.

MDM   Abbreviation for "master data management." MDM describes a system for the central harmonization and maintenance of master data in order to ensure data consistency across systems and applications. Through the central approach, redundant data retention is avoided.

MESA   The Manufacturing Enterprise Solution Association (MESA) is a U.S. industrial association that focuses on improving business processes in the manufacturing sector through the optimization of existing applications and the introduction of innovative information systems. Both the vertical and horizontal integration of information systems plays a significant role. Only shortly before the ISA, the MESA was the first organization to address the topic of MES in detail.

Minimum stock level   The minimum stock level is the stock in a warehouse that marks the threshold for triggering a procurement process; when levels are below or equal to this level, the process is triggered. Other terms for this are *minimum inventory level*, *order point*, and *reorder level*.

MRP   Abbreviation for "material resource planning." An MRP run is normally carried out by operative order planning through an ERP system with the task of checking and/or reserving the material stock for the order pool. We also speak of *inventory disposition*. The MRP function also can be perceived as a partial function of operative planning within a qualified MES.

Multithreading   A software application can be created in several *threads* (independently running partial processes) in order to use the processor power better.

NC   Abbreviation for "numerical control." Describes the digital (numerical) control of machine tools (CNC machines) in manufacturing technology. When needed, the processing programs (NC programs) are loaded from one of the connected computers to the machine control with the aid of the DNC system.

Net requirements   Arises from the gross requirements minus the available stock of a resource. The available stock is the balance of the physical stock plus open purchase order quantity minus reserved stock.

ODA Abbreviation for "operation data acquisition." ODA contains a summary of performance data on the performance object "order." Who has produced what quantity at what time over what period of time and at what workplace is recorded. Also known as *production data acquisition* (PDA).

OEE Abbreviation for "overall equipment efficiency," a key performance indicator (KPI) for machines and systems. The OEE is defined as the product of *availability*, *performance rate*, and *quality rate* for a machine.

OLAP Abbreviation for "online analytical processing." OLAP is a statistical method for the multidimensional analysis of data. Suitable data structures should be included in the design stage of the data models for effective information management. OLAP tools also contain multivariate evaluation functions.

OLE Abbreviation for "object linking and embedding." A technique based on COM technology for communication and data exchange between Windows applications, which allows for the embedding of objects of an application in other applications.

OPC Abbreviation for "openness, productivity, and collaboration" (previously for "OLE for process control"). A standardized software interface that enables data exchange between software ports from applications by different manufacturers. Originally, OPC was based on DCOM. The most recently released specification, OPC UA (unified architecture), unites all previous specifications independent of platform (without DCOM technology). The core of this specification describes a service-oriented architecture with Web services and thus follows the current trend in IT.

Order tracing In various directives (ISO, EN, etc.), complete documentation of processes is required; that is, all performance data from all work steps must be documented, and it must be possible to trace it via a tracing function. On the basis of a product ID (e.g., serial number) for the end product, it must be possible to ascertain all relevant components and processes retroactively.

PAA Abbreviation for "part average analysis." PAA is a method for the early recognition of stochastic and systematic errors in the value-creation process.

PCS Abbreviation for "process control system." A process control system controls and monitors in particular process-oriented applications.

Performance rate/performance level The performance rate is defined as the ratio between the quantity actually produced per time unit (e.g., pieces per hour) and theoretically possible quantity per time unit. It records idle time and exceeding of planned rate and processing times on a machine/equipment immediately. The performance rate is similar to the level of utilization.

Personnel management See *"HRM (human resource management)"*

Personnel timekeeping records In timekeeping, a distinction is made between attendance recording within the scope of an access control system and order-specific timekeeping records. Order-specific timekeeping records in conjunction with shift and wage models form the basis for wage accounting.

PLC Abbreviation for "programmable logic controller." Programmable control system for the production area that is distinguished by its robustness, reliability, and ability to run in real time.

PLM Abbreviation for "product lifecycle management." PLM is a holistic concept for IT-supported management of all product data beginning with design and engineering and throughout the entire lifecycle (change management) to scrapping.

ppm Abbreviation for "parts per million." Indicator for the quality of a production process that indicates how many elements from a quantity of 1 million fulfill a particular criterion.

PPS Abbreviation for "production planning and scheduling." PPS systems were the predecessors to MES and performed important tasks.

Process capability The capability of a process is measured using capability indices. The extent to which the measured results of selected characteristics lie within permitted tolerances is checked. If the sixfold distribution of the measured values is within the tolerance or action limits to the right and left of the average, the process is deemed to be capable (6Sigma method).

Product accompanying card/goods issue slip A goods issue slip accompanies a production unit and contains the data necessary for identifying the order, such as order number, article, quantity unit, and quantity. These documents generally have only limited validity within a production area. The MES must be able to create and print these documents when an order is released and also must be able to reproduce them if lost.

Product data/product definition management The main product definition data are the work plan and parts list. Both are core components of the MES. Therefore, management of the product definition data should be carried out in the MES.

Production control Production control releases production orders for production, establishes individual capacities at short notice (order distribution), tracks production processes, and ensures the timely delivery of the production orders. The sequence of the orders should be chosen so that the equipment and personnel capacities are used fully and as evenly as possible in as short a cycle time as possible. The current availability situation, which is controlled by disposition, should be considered.

Production depth The production depth is a measure of the number of production levels in the company. If only prefabricated components are assembled, the production depth is said to be low—if all components are produced by the company, the production depth is said to be high.

Glossary

Production logistics   Production logistics controls the material flow within the value-creation chain with regard to previous and subsequent relationships.

Production unit   The production unit (not to be confused with the concept from the ISA level model) is a defined quantity of an article that is independent of the actual order. Examples here are a lot (in series production) or a batch or charge (in process technology). A production unit can contain several orders, and an order can extend across several production units. The production unit receives an order-independent identifier (e.g., lot number, serial number, charge number, or unique article ID) that is linked with the connected order(s). Thus it serves in particular as a basis for product tracking and for recording all production-specific data.

Pull principle   The pull principle means that a point of consumption requests necessary material and preliminary products when they are needed ("pulling" from production). This principle is a core element of the *kanban* method (see also *Kanban*).

Quality management   Quality management organizes monitoring of the production process with test planning, test data collection, and monitoring of the test parameters. In addition to a suitable organization, integration of personnel and their constant further training, a CAQ system generally is a significant aid. See also *CAQ, SPC, SQC,* and *TQM*.

Quality rate   The quality rate is defined as the ratio between the usable quantity produced and the overall quantity produced. Thus it includes production units that are lost through scrapping or reworks.

Real-time cost control   When the provision of services is being recorded, the direct costs for material and time consumption are determined immediately and used for an instant comparison of target and actual costs. This monitoring is part of an early-warning system.

Recipe   A recipe includes parameters and raw materials that are involved in the functional or process technology and the relevant processing instructions. The recipe defines the ratio of the raw materials and a sequence program that controls the process.

Resource management   Resource management involves the definition and management of individual resources such as materials, machines, personnel, equipment, and means of transport, as well as their planning and provision for the various operations. Managing the production resources is one of the tasks of the MES.

RFC   Abbreviation for "remote function call." Special form of RPC defined by SAP (see *RPC*).

RFID   Abbreviation for "radiofrequency identification." Automatic identification of objects with high-frequency signals and transponders (data media). In an MES, RFID technology is used especially for controlling and tracing production units.

RPC Abbreviation for "remote procedure call." Communication between software processes through calling up a *remote* (i.e., located on a different computer) procedure.

SCADA Abbreviation for "supervisory control and data acquisition." System for machine-based use and monitoring that can take on partial tasks of an MES locally or can form an interface between machine control and an MES.

Scheduling Scheduling of the production orders (planning of dates) occurs in the MES with the aid of an operative planning system on the basis of the work plans for the article. A distinction is made between backward (starting with the latest delivery date) and forward scheduling (starting with the earliest production start), as well as scheduling of a bottleneck resource (bottleneck scheduling).

SCM Abbreviation for "supply-chain management." Systems for the management and monitoring of the global supply chain of a company. SCM systems manage both external and internal supply chains. One example of a partly external, partly internal supply chain is the provision of the material for the first work process of a product.

6Sigma 6Sigma is a statistical quality target (standard deviation based on freedom from defects) and also the name of a quality management method with the goal of zero-defect production. The name 6Sigma comes from the requirement that the next tolerance limit set should be at least 6 standard deviations from the average (6Sigma level). Only when this requirement is met is it possible to assume that a practically zero-defect production is being achieved, meaning that tolerance limits are hardly ever exceeded.

SOA/SOAP Abbreviation for "service-oriented architecture." The basic idea of service-oriented architecture is to categorize the business processes into individual services. The client calls up a service for a defined task (order to service), the server then processes this order, and the result (response from server) is passed back to the client. The best established technological approach to implementing SOA takes the form of Web services. The W3C (World Wide Web Consortium) has carried out an extensive standardization of Web services and data exchange via SOAP (protocol for data exchange via HTTP and TCP/IP), thus facilitating application of the technology in heterogeneous environments (see also *WSDL*).

SPC Abbreviation for "statistical process control." In SPC, individual process parameters are checked for their process capability. In particular, statistical methods are used to check whether a significant change arises in the location, dispersal, and/or distribution of the process. SPC is carried out on the basis of data measured online in the production process. Unlike SPC, SQC provides a statistical check for finished parts (see also *CAQ, Cp/Cpk, Quality management, SQC,* and *TQM*).

SQC Abbreviation for "statistical quality control." For the statistical quality control of the article, the generally manually collected data (random testing) are evaluated statistically and are made transparent in graph form in what are known as *rule cards* (e.g., XbarS cards), histograms, etc. Data are recorded separately for each production unit and distinguished between variable and attributive features (see also *CAQ, Cp/Cpk, Quality management, SPC,* and *TQM*).

SQL   Abbreviation for "Structured Query Language." Programming language for queries and data processing from a relational database.

SRM   Abbreviation for "supplier relationship management." Software system for definition, management, and control of supplier relationships of a company. SRM is located in the company management level as an independent system or forms an integrated part of an ERP system.

STEP   Abbreviation for "Standard for the Exchange of Product Data." STEP is a standard for the description and exchange of product information (see also *Product data/product definition management*) and can, for example, be used for data exchange among PLM systems or between a PLM system and an MES.

Supplier rating   The assessment of a supplier, for example, in DIN EN ISO 9001:2000 certification, can take place within an audit or on the basis of documents. Possible criteria are, for example, delivery reliability, product quality, credit worthiness, and QM system.

SVG   Abbreviation for "scalable vector graphics." SVG is a vector format based on XML for describing two-dimensional graphics. The format also supports the animation of objects whose characteristics can be manipulated via scripts. SVG allows for the integration of vector-based graphs in Web solutions.

SWF   Abbreviation for "Small Web Format/Shock Wave Flash." File extension for data in Flash format.

TCO   Abbreviation for "total cost of ownership." Consideration of the overall costs of a system. In addition to procurement costs, it includes running costs for maintenance and operations throughout the planned running time.

TCP/IP   Abbreviation for "Transmission Control Protocol/Internet Protocol." Best known and most often used protocol for communication on the Internet. TCP/IP is also used increasingly as a protocol on the automation and field level (for the connection of decentralized controls and sensors/actuators).

Test plan   In the test plan, all features to be checked in a product in order to be able to evaluate it are listed. A distinction is made between variable and attributive features. Variable features are measurement values with a target value and tolerances; for attributive features, an assessment (e.g., OK, not OK, grade, etc.) is given.

Glossary

TPM Abbreviation for "total productive maintenance." TPM is a method for the overall maintenance management of a producing company, including the required software systems, and is used for the continual improvement of production. TPM is often interpreted as total productive management in the sense of a comprehensive production system.

TPS Abbreviation for "Toyota production system." TPS was developed by Toyota in the 1950s and includes the following measures for increasing production efficiency: avoiding and reducing sources of loss, synchronizing process sequences, avoiding errors, improving machine productivity, and ongoing training and additional training of employees.

TQM Abbreviation for "total quality management." TQM includes all measures in management and in production that contribute to zero-defect production. See also *CAQ, Quality management, SPC, SQC,* and *TPM*.

UDDI Abbreviation for "Universal Description Discovery and Integration." Protocol for the publication and discovery of metadata (description) for Web services. Can be used for development and processing time (integration).

Value creation Indicator for the provision of a company's own services. In the production process, this is the value created in the individual operations.

W3C Abbreviation for "World Wide Web Consortium." Body for the standardization of technologies related to the World Wide Web.

WIP Abbreviation for "work in process." Recording and management of the stock found in production (semifinished products) as a partial function of an MES.

Workflow Organizing work processes by describing and determining definable processes and task-sharing processes that must be carried out in a defined sequence (parallel or serial).

WPF Abbreviation for "Windows Presentation Foundation." Programming interface (API) for user interfaces in Windows operating systems.

WSDL Abbreviation for "Web Service Description Language." A WSDL file describes a Web service and allows it to be used via any Web service client. See also *SOA/SOAP*.

XAML Abbreviation for "eXtensible Application Markup Language." Language for the description and creation of user interfaces of Windows Presentation Foundation (WPF) in XML.

XML Abbreviation for "eXtensible Markup Language." XML is a text-based metalanguage for describing hierarchically structured data, processes, etc. that has been standardized by the W3C.

Index

A

accompanying documentation, 77
active directory, 143
activity-based costing, 15, 23
ad hoc reports, 149
ADO, 131
advanced planning and scheduling (APS), 85
AJAX, 148
alarm management, 119
alarms and events, 137
analysis of the actual situation, 178
application-centered systems, 121
application server, 123
application support, 194
archiving, 133
article, 57
article groups, 57
asynchronous JavaScript and XML, 148
availability, 116

B

basic data, 59, 78
big bang, 173
bottleneck planning, 92
business logic, 124
by-products, 109

C

CAQ, 6
change management, 66
changes in working conditions, 175
changing tools, 103
collaborative production management (CPM), 1, 18
COM, 137
competency, 175
compliance, 25
compliance management, 25
compliance standards, 117
computer-aided design (CAD), 6
computer-integrated manufacturing (CIM), 6
continual improvement process (CIP), 16
continuous manufacturing, 197
contract, 186
contract specifications, 180, 186
control center, 181
control circuit, 177
cost control, 53, 110
cost control management, 23
creation of variants in the operation, 63
customer relationship management (CRM), 9
customer satisfaction, 185
customized software, 171
customizing, 186, 193

D

data warehouse, 14, 88, 149
database, 132
　interface, 142
　resource monitoring, 130
　running maintenance, 134
　scaling, 132
　several instances, 131, 133
database-centered systems, 121
degree of performance, 116
degree of quality, 116
demand planning, 85
dependency analyses, 118
digital factory, 6
digital mock-up, 6, 10, 219
DIN EN ISO 9001:2000, 26
discrete manufacturing, 197
distributed COM, 137
DMAIC, 28, 111, 159
DNC, 104
dynamic creation of master data, 57

Index

E

e-mail, 144, 151
early warning systems, 22
earnings, 175
employee motivation, 185
engineering component (EC), 44
enterprise production management (EPM), 2
enterprise resource planning (ERP), 9
escalation management, 24
ethernet TCP/IP, 139
event management, 22, 117
eXtensible Application Markup Language, 148

F

fat client, 147
FDA 21 CFR Part 11, 26, 36, 105
file interface, 142
first-level application support, 194
5S method, 157
Food and Drug Administration, 36
forward planning, 92

G

Gantt diagram, 91
general goals, 186
good manufacturing practices (GMP), 37

H

hardware/infrastructure operations, 194
Heijunka, 85
historical data access, 137
HMI, 44, 136
hotline support, 195
HTTP, 130
human resources management (HRM), 10

I

IEC, 35
implementation process, 191
implementation strategies, 173
importing master data, 56
incoming goods, 87
indirect value creation, 182
information distribution, 150
information management, 106
infrastructure, 179
input, 60
interactive control station, 91
interface to ERP, 140
interface to the production, 134
internal services, 177
interoperability, 43
ISA, 31, 54

ISA level model, 9
ISA S88, 11, 32
ISA S95, 33
isolated solutions, 180
IT infrastructure, 180

J

job security, 175

K

Kanban, 105
key performance indicator (KPI), 17, 24, 81, 116, 181

L

labor information management system (LIMS), 198, 199
labor relations, 175
LDAP server, 143
lean manufacturing, 27, 160
lean production, 160, 185
lean Sigma, 27, 28
lean 6Sigma, 28
long-term archive, 133

M

machine data, 79
machine data capture, 181
machine settings, 104
maintenance, 116, 194
  management, 118
  preventative, 118
managing production bin, 106
manufacturing execution system (MES), 11, 13
market situation, 187
master data management (MDM), 14, 20, 143, 215
material, 73
material disposition, 86
material flow control, 106, 110
material management, 181
material overhead, 60
material release list, 74
material usage, 74, 80
material usage function, 106
material warehousing costs, 88
MDA, 5
means of transport, 72
measuring equipment, 73
MESA, 38
mobile computing, 146
mobile solution, 146
Modbus TCP, 138
moving targets, 182

MRP, 5
MRP run, 34
multilingualism, 56
multivariate statistics, 22

N

NA 94, 38
NAMUR, 37
NE 33, 37
.NET framework, 148
.NET-3.0-API, 148

O

ODBC, 131
OEE, 116, 183
OLE for process control, 137
OPC A/E, 137
OPC command, 138
OPC DA, 137
OPC DX, 138
OPC HDA, 137
OPC UA, 138
OPC XML DA, 137, 138
openness, productivity, and
    collaboration (OPC), 136
operating concept, 193
operating data acquisition (ODA), 109
operating data capture, 181
operating resources, 62
operation, 58
optimal sequence, 85
order data, 79
order data management, 83
order fulfillment, 99
order planning, 94
order processing, 53
order time, 61
order tracing, 185
output, 59, 107
overhead, 60

P

parameterization options, 127
patches, 143
patchwork, 13
performance analysis, 117
performance data, 112
performance measurement, 155
personal digital assistant (PDA), 5, 146
personnel time, 81
personnel time recording, 81
planning algorithm, 91
planning costs, 65
platform independence, 124
PPS, 41
preventative maintenance, 118

proactive communication, 25
problems during implementation, 174
process, 110
process assurance, 110
process capability index, 111
process instructions, 74
process-oriented component, 44
process-oriented manufacturing, 197
process performance index, 111
process supervision, 111
process visualization, 136
product data integration, 19
product data management (PDM), 47, 85
product definition, 57
product definition data, 20
product definition management, 11, 19
product history, 66
product lifecycle management (PLM),
    6, 8, 13, 19, 47
product management system (PMS), 19
production, 109
production data acquisition, 5
production data integration, 21
production flow-oriented design, 53
production flow-oriented planning, 53, 83
production personnel, 71
production time, 61
productive time, 116
Profinet IO, 139
programmable logic controller, 13
project management, 191
project manager, 177
project plan, 185, 192
pull principle, 157
purchased products, 61

Q

qualification, 175
quality assurance, 110
quality data, 80
quality management, 181
quality overhead, 60

R

ratio of quantity to next sequence, 64
raw material, 61
real time, 22
real-time ability, 43
real-time data management, 21
real-time performance control, 117
recipe, 61
redundancy, 43
reel production, 109
reel transformation, 109
release management, 15, 190, 194
release of orders, 96

releases, 194
remote function call (RFC), 142
remote procedure call (RPC), 143
reporting, 148, 150
repository, 149
requirements, 186
resources management, 11
return on investment (ROI), 160
reverse planning, 92
rework, 110
RFID technology, 108
rich client, 147

S

scalability, 125
scalable vector graphics (SVG), 148
second- and third-level application support, 194
sequence planning, 91
service-oriented architecture (SOA), 18, 87, 129, 138
service overhead, 60
setup, 105, 109
setup time, 61
shock wave flash (SWF), 148
Short Message Service (SMS), 144, 151
sigma level, 167
Silverlight, 148
Simple Network Management Protocol (SNMP), 144
simulation, 190
6Sigma, 28, 29, 111, 159, 185
smartphone, 146
SOAP, 130, 138
SPC, 102
specifications, 186, 192
SQC, 102
standard software, 171
standard user interfaces, 145
standardization, 7, 135
storage costs, 60
style concept, 145
suitability for updates, 127
supervisory control and data acquisition (SCADA), 13, 45, 104, 136
supply-chain management (SCM), 10
system administrator, 193
system data, 77

T

target management, 16
task schedule, 11
TCP/IP, 129
terminal, 70

test run, 105
thin client, 147
time container, 93
tools, 72
total cost of ownership (TCO), 190
total productive maintenance (TPM), 118
Toyota production system (TPS), 27, 157
traceability, 110
tracking and tracing, 12, 181
training management, 192
transformation, 60

U

unified architecture, 138
updates, 143, 194
usage and visualization, 144
user interfaces, customer-specific, 128

V

value-stream mapping, 180
variant, 58
VDA, 39
VDI, 36
VDMA, 40
visualization, 145

W

warehousing costs, 94
Web Service Description Language (WSDL), 130
web services, 142
work instructions, 74
work intensity, 175, 176
work plan, 53
  updated, 89
work scheduling, 89
work sequence, 59
worker information, 75
workers, 193
working conditions, 175
working time regulations, 175
World Wide Web Consortium (W3C), 129
WPF, 148

X

XAML, 148

Z

zero-defect production, 29, 118
ZVEI, 40

CPSIA information can be obtained at www.ICGtesting.com
Printed in the USA
LVOW10*2314160316

479474LV00009B/38/P